TRANSDUCERS
FOR
AUTOMATION

TRANSDUCERS FOR AUTOMATION

MICHAEL F. HORDESKI, P.E.

Control Systems Consultant
Siltran Digital
Atascadero, California

and

Lecturer
California Polytechnic State University
San Luis Obispo, California

VNR VAN NOSTRAND REINHOLD COMPANY
New York

Copyright © 1987 by Van Nostrand Reinhold Company Inc.
Library of Congress Catalog Card Number 86-15674
ISBN 0-442-23700-6

All rights reserved. No part of this work covered by the copyright hereon
may be reproduced or used in any form or by any means—graphic, electronic,
or mechanical, including photocopying, recording, taping, or information
storage and retrieval systems—without written permission of the publisher.

Printed in the United States of America

Van Nostrand Reinhold Company Inc.
115 Fifth Avenue
New York, New York 10003

Van Nostrand Reinhold Company Limited
Molly Millars Lane
Wokingham, Berkshire RG11 2PY, England

Van Nostrand Reinhold
480 La Trobe Street
Melbourne, Victoria 3000, Australia

Macmillan of Canada
Division of Canada Publishing Corporation
164 Commander Boulevard
Agincourt, Ontario M1S 3C7, Canada

16 15 14 13 12 11 10 9 8 7 6 5 4 3 2 1

Library of Congress Cataloging-in-Publication Data

Hordeski, Michael F.
    Transducers for automation.

    Bibliography: p.
    Includes index.
    1. Transducers. 2. Automatic control. I. Title.
TJ223.T7H67   1987   629.8'043   86-15674
ISBN 0-442-23700-6

To students, young and old

For the first time in thousands of years, we face again a situation that can be compared with what our remote ancestors faced at the time of the irrigation civilization. It is not only the speed of technological change that creates a revolution, it is its scope as well. Above all, today, as seven thousand years ago, technological developments from a great many areas are growing together to create a new human environment. This has not been true of any period between the first technological revolution and the technological revolution that got under way two hundred years ago and has still clearly not run its course.

Peter F. Drucker

PREFACE

Factory automation is quickly becoming one of the most important areas of technology. Its emergence has been fueled by increased international competition, inflation and rapidly changing price factors, the high cost of capital, the decreasing availability of skilled labor, and the increasing emphasis on quality.

Manufacturers of goods of all kinds are driven to automate using transducers of all types and shapes. Dramatic changes in computing power allow lower implementation costs while enhancing the value of the actual equipment required. This provides an increased capacity for factory automation at all levels of manufacturing.

As these trends continue to enhance the return of investment potential to industry, the interest in factory automation transducers and technology can be expected to intensify. Manufacturers who do not utilize factory automation techniques are expected to find it difficult to survive in the late '80s and almost impossible to compete in the '90s.

Business may continue to grow in many areas that had been depressed, but the companies that will do best are those that utilize automated manufacturing equipment and techniques to produce less expensive products of higher quality.

An understanding of factory automation transducers is required for those involved in the manufacturing and use of automated equipment. This includes the users and suppliers of much processing equipment.

The pressure for continued education in design and manufacturing is intense. Some estimates indicate that over two-thirds of the one and a half million U.S. engineers obtained their degrees when calculations were done with a slide rule. Since training and education directly affect productivity, most real growth will depend on new technology.

This book provides the key ingredients from two vital areas that are required in modern factory systems: the transducers that are used to sense and monitor discrete and continuous processes, and the techniques that must be considered for the rapidly growing field of automation.

The book is written in nontechnical language, but the treatment provides enough depth to be useful to practicing engineers and engineering faculty in many educational institutions. The departments that usually offer these courses are Electrical Engineering, Engineering Technology, and Mechanical Engineering. This book was developed in part during my transducer courses at the California State Polytechnic University in San Luis Obispo. I wish to thank those at other institutions with active programs in this technology including George Washington University, Iowa State University, MIT, Texas A&M, UCLA, University of Michigan, Princeton, Stanford, University of Illinois, University of California, University of South Florida, USC, and the University of Wisconsin. These schools all have strong programs in industrial instrumentation and have contributed to the advance of education in this important area. I also wish to thank Fred Sandritter of AMETEK, Greg Chambers of Rexford, Martin Conway of Volumetrics, and Michael Scelzo of Panametrics for their support in this project. A special note of thanks is deserved by Dee at Siltran Digital, who typed and proofread the entire manuscript and diligently kept the project moving according to schedule.

Chapter 1 serves as an introduction and covers such subjects as transducer specifications, test procedures, accuracy limitations, control system terminology, measurement fundamentals, multiplexing, and the treatment of noise and interference.

Temperature measurement is the subject of Chapter 2. Temperature is the most widely measured variable. It is used as a direct indication of a process as well as to infer other conditions that must be monitored for automation. This chapter covers resistance temperature detectors, thermocouples, thermistors, semiconductor methods, radiation techniques, and methods for reducing errors in temperature measurement.

Pressure transducers are of interest since many materials in processing are in a liquid or gaseous form during all or part of production. The measurement of pressure is fundamental in the automation of these operations. Chapter 3 considers inductive pressure transducers, piezoelectric sensors, capacitive transducers, and the different types of strain gauge pressure transducers.

Flow measurement is the subject of Chapter 4, which covers differential pressure methods, turbine flowmeters, variable-area meters, fluid characteristic sensors, electromagnetic flowmeters, sonic and ultrasonic flowmeters, weirs and flumes, and positive displacement meters.

Density and weight measurement are considered in Chapter 5. Density may be measured with angular position, displacement, fluid dynamics, capacitance, hydrometric, hydrostatic, oscillating-Coriolis, radiation, and vibration techniques. Weight measurement may use lever scales, load cells, strain gauges, nuclear radiation, and magnetostrictive sensing.

Viscosity, moisture, and humidity measurements are the subjects of Chapter 6. Viscosity can be measured using bubble timing, capillary techniques, falling pistons, or rotational and vibration methods. Moisture and humidity can be measured with hygrometric or psychrometric techniques or with dew-point sensing methods.

Level detection is considered in Chapter 7. The applicable sensing methods include bubblers, capacitance probes, antenna sensors, differential pressure methods, conductivity probes, diaphragm sensors, displacement techniques, floats, impedance probes, optical methods, resistance tapes, radiation techniques, rotating-paddle switches, thermal sensors, vibrating-reed switches, and ultrasonic sensors.

Chapter 8 is concerned with displacement and proximity measurement for position, velocity, and acceleration applications. Displacement and proximity techniques include resistance, capacitive, inductive, and optical sensors. This chapter also includes a discussion of linear velocity measurement, tachometers, and accelerometers.

Chapter 9 covers a number of miscellaneous measurements including turbidity, thermal and electrical conductivity, pH, flame and heat sensors, leak detectors, metal detectors, noise, torque, and stress and strain.

Automation fundamentals are the subject of Chapter 10. Topics include computer-aided systems, factory-related software, factory software interfaces, data processing and communications networks, industrial local area networks (LANs), robots, CAD/CAM, computer-integrated manufacturing (CIM), positioning systems, process control, multiplex configuration, programmable controllers, laser machining, numerical control, software selection, and distributed processing and distributed control.

The present aims in factory automation are directed at developing unified systems for directing the activities of interconnected groups of robots and machines. This activity will pave the way for totally automated factories. Total automation does not mean a factory without people but rather one automated to the fullest practical extent.

Most factory systems in the future will be made up of workers teamed up with robots and other automatic machines. How well these systems operate will depend to a large extent upon the expertise that went into the selection and implementation of the technology.

MICHAEL HORDESKI, P.E.
Atascadero, California

CONTENTS

Preface/ix

1. Introduction/1

The Concept of Measurement in Automation Applications/1   An Introduction to Transducers/4   Typical Transducer Test Procedures/6   Accuracy/11   System Accuracy/12   Control System Concepts/14   Feedback Control/15   Feedforward Control/16   System Considerations/16   Converter Application/20   Multiplexing/22   Noise and Interference/25   Exercises/29

2. Temperature Measurement/30

Introduction/30   Resistance Temperature Detectors/30   Platinum Resistance Sensors/32   Thermistors/33   Excitation Methods/34   Transistors and Integrated Circuits/35   Thermocouples/36   Radiation Techniques/40   Reducing Errors in Temperature Measurement/44   Exercises/48

3. Pressure Transducers/50

Pressure Measurement/50   Inductive Pressure Transducers/50   Piezoelectric Pressure Transducers/54   Capacitive Pressure Transducers/56   Potentiometric Pressure Transducers/57   Strain Gauge Pressure Transducers/59   Pressure Transducer Errors/65   Exercises/69

4. Flow Transducers/71

Introduction to Differential Pressure Flow Transducers/71   Orifice Plates/71   Venturi Tubes and Flow Nozzles/73   Turbine Flowmeters/75   Variable-Area Meters/77   Fluid Characteristic Sensors/80   Electromagnetic Flowmeters/84   Sonic and Ultrasonic Flowmeters/85   Weirs and Flumes/86   Positive Displacement Meters/88   Mass Flowmeters/89   Exercises/91

5. Density and Weight Measurement/93

Density/93   Gas Density Detectors/95   Angular Position Liquid Density Sensors/95   Ball Liquid Density Sensor/96   Capacitance Liquid Density Meter/97   Displacement Liquid Density Sensor/97   Displacer Float Density Sensor/98   Electromagnetic Suspension Density Sensor/99   Fluid Dynamic Liquid Density Sensor/100   Hydrometers for Liquid Density Measurement/101 Hydrostatic Head Devices for Liquid Density/101 Oscillating-Coriolis Liquid Density Sensor/102 Radiation Liquid Density Sensor/103 Sound Velocity Liquid Density Sensor/105 Torsional Vibration Liquid Density Sensor/106 Vibrating-Plate Liquid Density Sensor/106 Vibrating-Spool Liquid Density Sensor/107 Vibrating-Tube Liquid Density Sensor/108 Weight-Bulb Density Sensor/109 U-Tube Density Sensor/110 Straight-Tube Density Transmitter/110 Direct Density Controller/111 Gas Density Sensors for Operational Conditions/111 Displacement Gas Density Sensor/111 Centrifugal Gas Density Sensor/112 Application and Limitations/112 Weight Sensors: Application and Selection Considerations/113 Spring Balance Scales/114 Mechanical Lever Scales/114 Hydraulic Load Cells/116 Hydraulic Totalizers/118 Pneumatic Load Cells/118 Strain Gauge Load Cells/119 Semiconductor Strain Gauges/121 Semiconductor Strain Gauge Load Cells/121 Nuclear Radiation Sensors/121 Inductive and Reluctance Load Cells/122 Inductive Sensing Techniques/122 Variable-Reluctance Techniques/123 Magnetostrictive Sensing Techniques/123 Beam-Type Strain Gauge Transducers/124 Monorail Weighing Transducers/125 The Direct Weighing of Tank Legs/125 High-Temperature Load Cells/126 Exercises/126

6. Viscosity Measurement/128

Pressure Drop Measurement Techniques/128 Oscillation Techniques/131 Torque and Weight Techniques/132 Moisture and Humidity Measurements/135 Hygrometric Techniques/136 Psychrometric Techniques/139 Dew-Point Sensing Techniques/140 Humidity Measurement Errors/142 Humidity and Moisture Automation Considerations/146 Exercises/148

7. Level Measurement/150

Applications/150 Antenna Level Sensors/151 Bubbler Systems/152 Capacitance Probes/154 Conductivity Probes/156 Diaphragm Level Detectors/156 Differential Pressure Level Detectors/157 Impedance Probes/158 Level Measurement Using Displacement Techniques/159 Level Gauges/164 Optical Level Switches/165 Radiation Level Sensors/167 Resistance Tapes/173 Rotating-Paddle Switches/174 Tape Level Devices/175 Inductively Coupled Float and Tape Detectors/176 Wire-guided Thermal Sensors/177 Surface Sensors/178 Vibrating-Reed Switches/179 Ultrasonic Level Detectors/180 Damped-Sensor Level Switches/181 On-Off Transmitter Level Switches/181 Continuous Level Detectors/182 Exercises/183

CONTENTS xv

8. Displacement and Proximity Measurement/184

Applications/184   Resistance Sensors/184   Capacitive Sensors/185
Inductive Sensors/186   Digital Measurement Techniques/187   Position
Transducer Considerations/190   Linear Velocity Measurement/191
Tachometers/192   Acceleration Transducers/196   Piezoelectric
Accelerometers/197   Accelerometer Measurement Systems/198
Exercises/200

9. Other Physical Measurements/202

Turbidity/202   Thermal Conductivity/203   Electrical Conductivity/205
Hydrogen Ion Concentration/207   Flame Sensors/209   Heat Sensors/
210   Electric-Conduction-type Detectors/210   Radiation Sensors/211
Leak Detectors/213   Metal Detectors/217   Noise Sensors/218
Environmental Considerations/223   Calibration/223   Stress and Strain
Measurement/224   Metallic Strain Gauges/224   Semiconductor Strain
Gauges/227   Strain Gauge Characteristics/228   Visual Techniques/232
Force and Torque Transducers/233   Piezoelectric Force
Transducers/234   Reluctive Force Transducers/235   Capacitive Force
Transducers/235   Torque Transducers/236   Exercises/239

10. Automation Fundamentals/241

Computer-Aided Techniques/241   Developing Factory Software
Interfaces/243   Interorganizational Communications/243   The Data
Model/245   Physical Data Structures/246   Application Development
Considerations/247   Application Software Characteristics/248
Structured Systems/248   Structured Design/248   Process Control/249
Multiplex Configuration/252   Process System Considerations/252
Positioning Control/253   Programmable Controllers/255   Robotic
Systems/256   The User of Laser Machine Tools/258   Numerical
Control/259   Software Selection/262   Sculptured Surfaces/264
Distributed Processing/265   Distributed Control/267   Computer
Networks/268   Factory Network Considerations/272   Language
Considerations/274   Network Protocol/275   HDLC Protocol/278
Addressing/280   Software Portability/281   MAP/283   Exercises/284

Bibliography/285

Index/297

1
INTRODUCTION

THE CONCEPT OF MEASUREMENT IN AUTOMATION APPLICATIONS

Measurement is defined as the extraction from physical and chemical systems or processes of signals which represent parameters or variables. The performance of an automation system can never surpass that of the associated measuring devices. A basic example is a human being. Information about the surroundings is acquired through the five senses, sight, smell, hearing, taste, and touch. This information is converted into electrical impulses and passed on to the proper section of the brain, which processes the information and sends a signal which causes a movement toward the objective. This is a closed-loop operation. Human performance deteriorates as the senses become impaired; our capability to drive an automobile decreases as our ability to see and hear decreases. Human performance can be improved if the senses are extended. Our ability to drive an automobile can be increased by the use of corrective glasses or a hearing aid.

The output of a measuring device is information concerning the state of the process to which it is connected. The information output may be relative to some previously defined point of reference.

A measuring instrument that has its output compared to a precision reference is called a standard. The standard is compared to an arbitrarily chosen reference of suitable magnitudes which is normally assumed to be unvarying.

Measurement engineering is a recognized discipline with the basic principle of valid measurement defined as follows:

A valid measurement is being made when the amount of information obtained is maximized and the amount of energy being taken from the process to obtain the measurement is minimized.

As an example of this, consider a measurement using a strain gauge. A stiff strain gauge can alter the stress field of the test piece. Normally, as energy is drawn from a process, the process is altered, and the measurement may not be strictly correct. Thus, a basic technique in making measurements is to minimize the alteration of the process, and thus reduce the effect on the measurement to a level of insignificance. When this technique is followed in most applications, the measurements are valid for most practical purposes, although some energy transfer is involved in all measuring systems.

"Transducer" or "sensor" is a general term for a sensing device in this book. The common mercury thermometer is a temperature sensing device which converts the temperature surrounding the mercury column into an equivalent length on a numbered scale. Sensing transducers can be classified as follows:

1. Self-generating transducers. These produce an output from a single energy input. A thermocouple is an example of this type of transducer or sensor.
2. External power transducers. These require energy inputs to provide an output. An example is a resistance thermometer, which requires an energy input to excite the resistor being observed.

The purpose of a measurement in an automation system often is to obtain as close an approximation as possible to the true value of a process quantity by comparing it to a standard or reference.

The complete measurement system can include a comparator, the reference, an amplifying unit, and the transducer or sensor. Errors are reduced with the proper application of the sensor and by the use of compensation techniques. The actual value of the measurement is a value of the process quantity.

The following definitions generally apply to all measuring devices including the transducers and sensors discussed in this book.

Accuracy—Conformity of an indicated value to an accepted standard value, or true value.
Repeatability—The closeness of agreement among a number of consecutive measurements of the output for the same value of input under identical operating conditions.
Resolution—The smallest interval between adjacent discrete details which can be distinguished from one another.
Hysteresis—The maximum difference obtained for the same input between upscale and downscale output values during a full-range transverse swing in each direction.
Sensitivity—The ratio of change in output to a change in input magnitude.
Reproducibility—The ability of a system of elements to maintain the same output/input precision over a period of time.

The definition of accuracy warrants additional discussion. We know that accuracy is a relationship to a true quantity, but to demonstrate its presence is not always easy. Absolute accuracy is unobtainable, but the basic definition of accuracy can be approached under highly controlled conditions such as those that exist at the National Bureau of Standards (NBS). The value of a measurement is based on a calibration standard with a known relationship and traceable to an NBS reference.

In a large number of automation applications, it is repeatability and not accuracy that is more critical. The use of transducers or sensors in these automation systems generally requires the consideration of the following:

1. What are the characteristics of the process that the transducer or sensor is required to measure?
2. How well will the measurement represent the actual characteristic or condition of the system?
3. What will the measurement indicate with regard to the actual process or manufacturing operation?
4. How is the instrument output eventually going to be used?

How well the measurement represents the true value of the measured quantity is affected by a number of factors. Temperature is one example. The main question is to decide if there is one measurable temperature in the process that is representative. Problem conditions such as stratification, pockets, and hot spots can result in significant differences in temperature for different measurement points. We try to eliminate these differences by carefully choosing the most representative measuring point, and to recognize that it may be less than absolutely representative.

Transducer accuracy is normally specified on a steady-state basis. Dynamic performance tells us how the measurement varies with respect to time. This is usually of equal importance. Again, consider a temperature measurement system. Thermal elements are usually installed in thermowells to protect them from the process fluids and to make withdrawal possible without interrupting process operation. The thermowell introduces a considerable time lag in the measurement system. A bare thermal element in a fluid stream can respond to a temperature change in a few seconds, while the same element in a thermowell and the same stream could require minutes to respond.

The outputs of transducer measurements on a process can be considered to be time-varying characteristics of product properties. There are three general classes of process output. The one that yields the most complete description of data is the analog representation, which is a point-by-point variation in time of each product property. Process information in such detail may be unnecessary, or even undesirable. The statistical parameters of mean and standard deviation are then used to describe a large quantity of process data. A third representation is based on the frequencies occurring in the input signal, and is called the power density spectrum.

At one time, instrumentation was considered as an added cost to the producing plant. Transducers are now becoming a major factor in reducing rather than increasing the overall production investment. One example is automatic custody transfer in the petroleum industry. Manual gauging has been replaced with automatic blending systems which greatly reduce tankage requirements as well as increasing accuracy. Modern measuring instruments can be designed and installed to provide better reliability, as well as increased accuracy, than the manual gauging and sampling techniques.

AN INTRODUCTION TO TRANSDUCERS

In this book we consider typical transducers or sensors that might be used in automation systems. For our purposes a transducer is any device that can be used to determine the value, quantity, or condition of some physical variable or phenomenon which must be monitored.

The purpose of the transducer is usually to measure the magnitude of some particular phenomenon for control purposes. The measurement consists of an information transfer with an accompanying energy transfer. Since energy cannot be withdrawn from a system without changing it in some way, we strive to keep the energy transfer small so that the measurement does not affect the quantity being measured. Transducers use a number of techniques to produce the information using this energy transfer.

Piezoelectric energy can be converted into an electrostatic charge or voltage when certain crystals are mechanically stressed. The stress can be from compression, tension, or bending forces. These can be exerted upon the crystal by a sensing element or by a mechanical member linked to the sensing element.

Energy can also be converted into a change in resistance of a semiconductive material due to the amount of illumination on the semiconductor surface. In some light-sensitive transducers the change in illumination may be controlled by a shutter or mask between a light source and the photoresistive material. The shutter may be mechanically linked to a physical sensing element such as a pressure diaphragm or a seismic mass.

Energy may also be converted into a change in voltage due to a junction between dissimilar materials being illuminated. This principle is utilized in some light sensing devices.

Transducers with a mechanical means to change the capacitive coupling between elements use energy which is converted into a change of capacitance. The capacitor consists of two conductors or plates separated by a dielectric. The change of capacitance occurs if a displacement of the sensing element causes one conductive surface to move relative to the other conductive surface. Some transducers use the moving plate as the sensing element; in others both plates are stationary and the capacitive charge occurs as the result of a change in dielectric.

Energy can also be converted into a voltage induced in a conductor by a change in magnetic flux. This type of transducer does not require any excitation and is self-generating. The change in magnetic flux can be accomplished by the movement of a magnetic material and the conductor. Energy can be converted into a change of the self-inductance of a coil due to a displacement of the coil's core. The core can be linked to a mechanical sensing element.

The three major factors that enter into the selection and installation of transducers and sensors in automation applications include the significance of the measurement, its proper application, and cost. Sound technical judgment based on these factors is required for the selection of necessary and appropriate measurements for efficient plant operation.

Selecting a transducer can involve the following considerations:

1. The purpose of the measurement.
2. The treatment of overloads before and during the time the measurements are taken.
3. The accuracy of the measurement.
4. The lower and upper limits of the frequency response needed.
5. The maximum error that can be tolerated during static conditions and during and after exposure to transient environmental conditions.
6. The limitations on excitation and output.
7. The power requirements.
8. The transducer's effect on the measurand.
9. The cycling time or operating life.
10. The failure modes of the transducer as well as the hazards due to a failure in a component in other portions of the system.
11. The human engineering requirements.
12. The environmental conditions.
13. The data transmission technique to be used.
14. The processing system to be used.
15. The type of data display to be used.
16. The accuracy and frequency response capabilities of the transmission, processing, and display systems.
17. The signal conditioning required.
18. The available transducer excitation voltage.
19. The current drawn from the excitation supply.
20. The load on the transmission circuit due to the transducer.
21. The filtering required in the transmission or data processing systems.
22. The requirements for the detection of and compensation for errors.

Transducer operation can be specified as continuous or cyclic, where the total number and the on-time or duration is stated.

The mechanical characteristics of the transducer may include the specifying of certain case materials and the exact nature of the sealing required for the environmental effects of corrosive or conducting fluids.

The most commonly used transducers are used to detect either temperature or pressure changes. Some temperature transducers may convert the thermal expansion of a liquid or solid into electrical signals through a mechanical linkage. Others may use resistors, diodes, thermistors, or thermocouples to sense temperature variations by monitoring changes in such parameters as conductivity or voltage. We are more concerned with this second class of transducers for most automation systems.

TYPICAL TRANSDUCER TEST PROCEDURES

The purpose of this section is to illustrate and clarify accuracy-related terms. It is intended to indicate generalized test methods. Test procedures are given for the following: accuracy, dead band, drift, hysteresis, linearity, repeatability, and reproducibility. The tests described here are for the determination of static performance characteristics, not dynamic characteristics. When relating performance characteristics such as values of accuracy to values of other terms such as linearity, hysteresis, dead band, and repeatability, equivalent units must be used.

The accuracy rating of a device relates the characteristics being tested to the tolerance allowed on the test device; for example, in the case of dead band,

Test device—allowed dead band 0.2%
Measuring device—preferred dead band 0.02%
Measuring device—minimum dead band 0.06%

Here the preferred accuracy rating of the measuring means is one-tenth or less that of the device under test. When the accuracy rating of the reference measuring means is between one-third and one-tenth that of the device under test, the accuracy rating of the reference measuring means should be accounted for.

The device under test and the associated test equipment must be allowed to stabilize under steady-state operating conditions. Those operating conditions which would influence the test should be observed and recorded.

The number of test points to determine the desired performance characteristic of a device should be distributed over the range (for examples of the use of the terms "range" and "span," see Table 1.1). They should include points at or near (within 10%) the lower and upper range values. There should not be less than five points and preferably more. The number and location of these test

Table 1.1. Illustrations of the use of the terms "range" and "span".

Typical ranges	Type of range	Range	Lower range value	Upper range value	Span
1. Thermocouple type K					
0 to 2000°F	Measured variable	0 to 2000°F	0°F	2000°F	2000°F
−0.68 to 44.90 mV	Measured signal	−0.68 to 44.90 mV	−0.68 mV	44.90 mV	45.58 mV
0 to 20 (× 100 = °F)	Scale	0 to 2000°F	0°F	2000°F	2000°F
2. Flowmeter					
0 to 1000 lb/h	Measured variable	0 to 1000 lb/h	0 lb/h	1000 lb/h	1000 lb/h
0 to 100 in. H_2O	Measured signal	0 to 100 in. H_2O	0 in. H_2O	100 in. H_2O	100 in. H_2O

points should be consistent with the degree of exactness desired and the characteristics being evaluated.

Before observations are recorded the device under test should be exercised by several full-range excursions in each direction. At each point being observed the input should be steady until the device under test becomes stabilized at an apparent final value.

The device under test should not be tapped or vibrated unless the performance characteristic being tested requires such action.

In a typical calibration cycle we must first precondition the test device and then observe and record output values for each desired input value for one full-range traverse in each direction starting near the midrange value. The final input should be approached from the same direction as the initial input and we should apply the input in such a way as not to overshoot the input value.

The calibration curve is normally prepared as a deviation curve. First, we determine the difference between each observed output value and its corresponding ideal output value. This difference is the deviation and can be expressed as a percentage of ideal output span. The deviation is plotted versus input or ideal output. Figure 1-1 shows the percent deviation plotted versus percent input. A positive deviation indicates that the observed output value is greater than the ideal output value.

Measured accuracy can be determined from the deviation values of a number of calibration cycles. It represents the greatest positive and negative deviation of the recorded values, in both upscale and downscale output swings, from the reference or zero-deviation line. Measured accuracy can be expressed as a plus and minus percentage of the ideal output span.

8 TRANSDUCERS FOR AUTOMATION

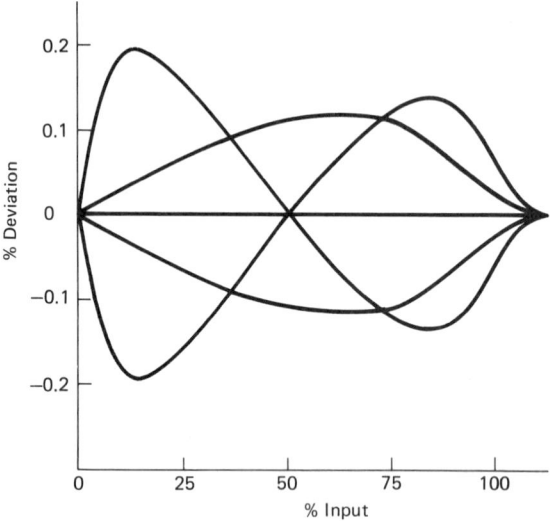

Figure 1-1. Typical calibration curves.

To determine the dead band, precondition the test device as discussed above then

1. Slowly increase or decrease the input to the device under test until a detectable output change is observed.
2. Record the input value.
3. Slowly vary the input in the opposite direction (decrease or increase) until a detectable output change is observed.
4. Record the input value.

The increment or difference between the two input values is the dead band. It should be determined from a number of cycles through steps 1 to 4. The maximum value is used. The dead band should always be determined at a number of points to make certain that the maximum dead band is found. The dead band may be expressed as a percentage of input span.

To find the point drift precondition the test transducer then

1. Adjust the input to the desired value without overshoot and record the output value.
2. Maintain a fixed input signal and fixed operating conditions for the desired time period.
3. At the end of the specified time interval record the output value.

In using the results of this test the dead band must either be negligible or such that it does not affect the drift value.

Point drift is the maximum change in recorded output value observed during the test period. It is expressed as a percentage of ideal output span for a specified time period.

Hysteresis results from the inelastic quality of an element or device. Its effect is combined with the effect of dead band. The sum of the two effects may be determined directly from the deviation values of a number of test cycles, and is the maximum difference between corresponding upscale and downscale outputs for any single test cycle. Hysteresis is then found by subtracting the value of dead band from the corresponding value of hysteresis plus dead band for a given input. The maximum difference is used. The difference may be expressed as a percentage of ideal output span.

Independent linearity can be found directly from the calibration curve, as follows:

1. Draw a deviation curve which is the average of corresponding upscale and downscale output readings.
2. Draw a straight line through the average deviation curve in such a way as to minimize the maximum deviation. It is not necessary that the straight line be horizontal or pass through the end points of the average deviation curve.

Independent linearity is the maximum deviation between the average deviation curve and the straight line. It is determined from the deviation plots of a number of calibration cycles. It is measured in terms of independent nonlinearity as a plus or minus percentage of ideal output span.

The average deviation curve is based on the average of corresponding upscale and downscale readings. This allows the independent linearity to be independent of the dead band or hysteresis. This assumes that if no hysteresis or dead band were present, the deviation curve would be a single line midway between the upscale and downscale curves.

Nonzero-based linearity may be found from the calibration curve using the following technique:

1. Plot a deviation curve which is the average of corresponding upscale and downscale output readings.
2. Draw a straight line such that it coincides with the average deviation curve at the upper range value and the lower range value.

Nonzero-based linearity is the maximum deviation between the average deviation curve and the straight line. It is found from the deviation plots of a number of calibration cycles. It can be expressed as a plus and minus percentage of ideal output span. Since the average deviation curve is based on the average of corresponding upscale and downscale readings it is independent of dead band or hysteresis. If no hysteresis or dead band is present, the deviation curve becomes a single line midway between the upscale and downscale readings.

Zero-based linearity can be found from the calibration curve using the following technique:

1. Plot a deviation curve which is the average of corresponding upscale and downscale output readings.
2. Draw a straight line such that it coincides with the average deviation curve at the lower range value or zero and minimizes the maximum deviation.

Zero-based linearity is the maximum deviation between the average deviation curve and the straight line. It is found from the plots of a number of calibration cycles. It is expressed as a plus or minus percentage of ideal output span. (*Example:* The zero-based linearity is $\pm 0.21\%$ of output span.)

The average deviation curve is based on the average of up and down readings. This allows the zero-based linearity to be independent of the dead band or hysteresis. Thus, if no hysteresis or dead band exists, the deviation curve becomes a single line midway between the upscale and downscale readings.

Repeatability can be found directly from the deviation values of a number of calibration cycles. It is the closeness of agreement among a number of consecutive measurements of the output for the same value of input approached from the same direction. Fixed operating conditions must be maintained. First, we find the maximum difference in percent deviation for all values of output considering upscale and downscale curves separately. The maximum value from either upscale or downscale curve is recorded. Repeatability is the maximum difference in percent deviation found above and is expressed as a percentage of output span.

Reproducibility can be found using the following technique:

1. Run through a number of calibration cycles.
2. Develop a calibration curve based on the maximum difference between all upscale and downscale readings for each input value. The deviation values are found from the calibration cycles performed in step 1.
3. Maintain the test device in its regular operating condition, energized and with an input signal applied.
4. At the end of a specified time repeat steps 1 and 2.

The test operating conditions can vary over the time interval between measurements providing they stay within the normal operating conditions of the test device. The above tests should all be performed under the same operating conditions.

Reproducibility is the maximum difference between recorded output values for a given input value. The maximum difference is always used. The difference is expressed as a percentage of output span per specified time interval.

ACCURACY

The subject of accuracy is often misunderstood. There are several reasons for this.

1. The term itself is poorly defined and widely misunderstood.
2. The interrelationships between accuracy, rangeability, calibration, and maintenance are not always recognized.
3. There is a tendency in the literature to use accuracy to present products in a more favorable light.

Accuracy is often defined as freedom from error or the absence of error. This is contrary to the widespread use of the term. When an accuracy statement is given as $\pm 1\%$ accuracy, in most cases what is meant is $\pm 1\%$ inaccuracy.

The purpose of any measurement is to obtain the true value of the quantity being measured, and error is thought of as the difference between the measured and the true quantity. Because it is impossible to measure a value without some uncertainty, it is equally impossible to know the exact error. All that can be stated in connection with the accuracy of a measurement, therefore, is the limits within which the true value will fall.

Accuracy-related terminology can be illustrated by a pattern of measurements, as shown in Figure 1-2. The spread in the measurements in the upper right-hand corner of the pattern represents the random error of the measurements. The deviation between the mean and the ideal (bull's eye) represents the systematic error. This error is repeatable and can be eliminated. Systematic error is also referred to as bias, which is the displacement of the measured or observed value from the true value.

The measurement in the lower left corner of the figure represents an illegitimate error which is caused by blunders, and such a false reading should be disregarded.

Accuracy of measurement is thus defined as the sum of random and systematic errors. If the purpose of an automation system is to maintain conditions at previous levels, without having an interest in the true values of these conditions, then it is desirable to reduce the random error, without changing the remaining bias. In such an application, a precise, repeatable measurement is more important than one based on absolute accuracy. If we are more interested in approach-

12 TRANSDUCERS FOR AUTOMATION

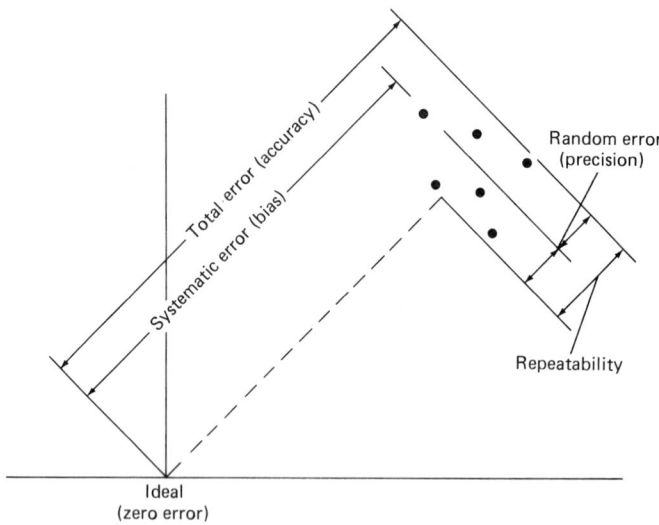

Figure 1-2. Accuracy-related measurements.

ing the true value of the measurement in order to serve such absolute purposes as materials accounting or quality control, then a repeatable detection is insufficient, and our attention must be concentrated on accuracy, which can be achieved only through the reduction of both random and systematic errors. If it is impossible to determine the systematic error, it can still be corrected by calibration against a fixed standard.

We can elaborate on accuracy statements beyond simply stating percentage values. Measurements can be given as the arithmetic mean of separate observations, taken in groups of successive runs at different times. The values have an estimated overall uncertainty, a standard error, and an allowance for possible systematic error.

SYSTEM ACCURACY

If there is no proven basis for evaluating the cumulative effect of component inaccuracies, then only an actual system calibration can reliably establish the total inaccuracy. The minimum number of components will result in the best accuracy for the total system. The exception to this is in digital systems where no additional error is introduced by the addition of functional modules.

Without an actual system calibration the evaluation of system accuracy must be based on assumptions. Table 1.2 shows system inaccuracies that can be expected under various conditions. The accumulated effect of component inaccuracies can be based on one of two assumptions:

INTRODUCTION 13

Table 1.2. System calibration.

% of full-scale output	Assumption 1	Assumption 2
20	±1.6%	±0.6%
50	±2.5%	±0.8%
80	±1.2%	±0.5%

1. The inaccuracy of each component is additive, and the system inaccuracy is the sum of component inaccuracies. This takes place in pure analog systems and in most practical systems represents a very conservative approach.
2. All component inaccuracies are neglected except for the least accurate component, and the total system inaccuracy is the same as the inaccuracy in this one component. This should only be used when the inaccuracy of this component dominates the known inaccuracies of the other components, since it is a very optimistic approach.

Figure 1-3 illustrates the inaccuracies of a system evaluated on a basis slightly more conservative than case 2 and much less conservative than case 1. It can be concluded that accuracy is not a single clearly defined number when one evaluates the performance of a multicomponent system under varying load conditions. We can, however, draw some qualitative conclusions:

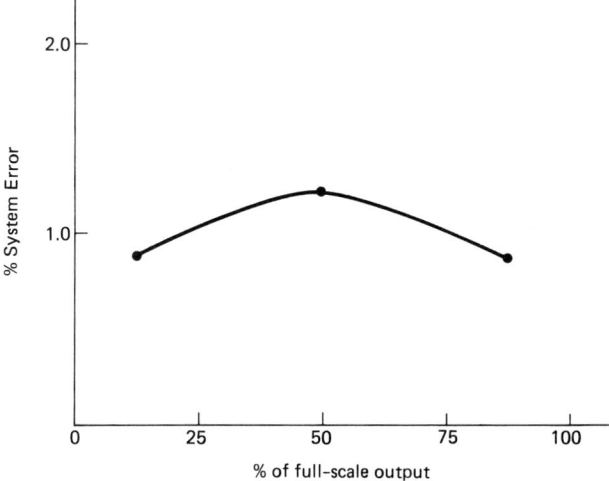

Figure 1-3. System inaccuracy curve.

1. Accuracy is likely to be improved by reducing the number of components in mixed systems of analog and digital components.
2. Accuracy is meaningful only in combination with rangeability. The wider the rangeability required, the more inaccurate the measurement is likely to be.

Linear analog system accuracies are less affected by rangeability requirements than are nonlinear analog systems, and the rangeability effect on digital systems is minimal.

In the area of calibrating sensors and systems, several points should be considered: (1) the accuracy of a multicomponent system is unknown unless calibrated as a system, (2) the calibration equipment must be at least three times more accurate than the system being calibrated, and (3) periodic recalibration is an essential prerequisite to good automation control.

CONTROL SYSTEM CONCEPTS

The development of automation systems has been equated in importance to the industrial revolution in the nineteenth century. In many respects the introduction of automatic control systems was a second industrial revolution. In the nineteenth century we learned to harness and use various forms of natural energy; in the twentieth century we have developed devices that can make the decisions necessary to control the various forms of energy.

The principles used in automatic control cut across virtually every scientific field. Today the basic principles of automatic control have a wide range of applications and interest. The real need for automatic control is obvious in some industries. In assembly line manufacturing facilities the need for automation is often critical. A machine in many cases is more suitable, for both economic and safety considerations, to perform the numerous machining and handling tasks involved.

The automation of a processing system might be viewed differently. Since most process equipment operates at a constant load, one might suggest that the best solution to the control problem is to set all the variables that affect the process to their proper positions and then let the process take over. The difficulty with this reasoning is that all the inputs to the systems can seldom be fixed. Most process systems are subject to many inputs. Changes in such variables result in disturbances in the process, unless some corrective action is taken.

In the water heater shown in Figure 1-4 the heater consists of a tank from which hot water is obtained by bubbling live steam directly into the tank, which is full of water. Cool water enters at the bottom of the tank and the hot water leaves at the top. A valve is used to regulate the flow rate of steam into the heater. In this system, if all other factors were constant, the temperature of the

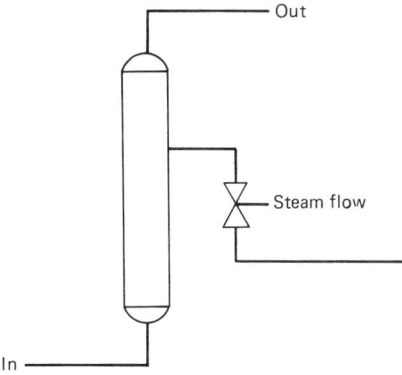

Figure 1-4. Water heater control.

outlet could be controlled by placing the steam valve at the proper setting. However, if the temperature of the inlet water changes, the outlet temperature will also change unless corrective action is taken. Other variables which affect the process are the flow rate of the water, the steam supply pressure, the steam quality, and the ambient temperature. A change in any one of these variables can cause a change in the water outlet temperature unless some corrections are made.

FEEDBACK CONTROL

The two concepts that form the basis for most automatic control strategies are feedback or closed-loop control and feedforward or open-loop control. Feedback control is the more commonly used technique and is the underlying concept on which much of today's automatic control theory is based.

Feedback control is a strategy designed to achieve and maintain a desired process condition by measuring the process condition, comparing the measured condition with the desired condition, and initiating some corrective action based on the difference between the desired and the actual condition.

The feedback strategy can be explained by the actions of a human operator attempting to control a process manually. Consider the procedure that one might use in the control of the hot water heater described earlier. The operator would read the temperature indicator in the hot water line and compare its value with the temperature desired. If the temperature were too high, one would reduce the steam flow, and if the temperature were too low, one would increase it. Using this strategy the operator would manipulate the steam valve until the error was eliminated.

An automatic feedback control system would operate in much the same manner. The temperature of the hot water is measured and a signal is fed back to a device which compares the measured temperature with the desired temperature. If an error exists, a signal is generated to change the value position in such a manner that the error is eliminated. The main difference between the manual and automatic means of controlling the heater is that the automatic controller is more accurate and consistent. Both systems contain the essential elements of a feedback control loop.

Feedback control has definite advantages over other techniques, such as feedforward control, in both the relative simplicity required and the potential for successful operation in the face of unknowns. It works well as a regulator to maintain a desired operating point by compensating for various disturbances which affect the system since it can initiate and follow any changes demanded in the operating point.

FEEDFORWARD CONTROL

Feedforward control is a basic method used to compensate for uncontrolled disturbances entering the system. In this method the control action is based on the state of a disturbance input without reference to the actual system condition. In theory, feedforward control yields much faster correction than feedback control, and in the ideal case compensation is applied in such a manner that the effect of the disturbance is never seen in the process output.

A skillful operator might use a simple feedforward strategy to compensate for changes in the inlet water temperature of a water heater. Detecting a change in inlet water temperature, one could increase or decrease the steam rate to counteract the detected change. The same compensation could be done automatically with an inlet temperature detector designed to initiate the appropriate corrective adjustment in the valve opening.

The concept of feedforward control is powerful, but it can be difficult to implement properly in many applications. In many cases disturbances cannot be accurately measured, and therefore feedforward concepts cannot be applied. Even in applications where all the inputs can be either measured or controlled, the correct action to be taken to compensate for a particular disturbance is not always known. In many actual applications, feedforward control is used in conjunction with feedback control in order to handle unknown contingencies which might otherwise disturb a pure feedforward control system.

SYSTEM CONSIDERATIONS

To accommodate an analog sensor voltage for analog-to-digital conversion, some form of scaling and offsetting may be required using an amplifier. To convert analog information from more than one source, additional converters or a multiplexer will be necessary. To increase the speed at which information may be accurately converted, a sample hold might be used. To compress analog signals, a logarithmic amplifier can be used.

The automation systems design should begin with the choice of sensors. If the systems design considers the selection of the transducers, this can go a long way toward easing the overall design task. In monitoring or controlling motor shafts, the automation design may have the choice of signals from at least three different position sensing approaches: shaft encoders, synchros, or potentiometers. Temperature measurements may be accomplished by thermocouples, thermistors, or RTDs, while force can be measured by strain gauges, or obtained by integrating the output from accelerometers.

If transducer signals must be scaled from millivolt levels to an A/D converter's typical ± 10-V full-scale input, an operational amplifier may be the best choice. If the system involves a number of sources, each transducer might be provided with a local amplifier so that the low-level signals are amplified before being transferred.

In some cases, the sum of the individual accuracy factors can result in a total error which exceeds the desired system accuracy. It may be necessary to reduce the total error to the required accuracy limits by one or more of the following methods:

1. Specifying the accuracy over a total error band rather than on an individual parameter basis.
2. Using in-system calibration techniques with corrections performed by data reduction.
3. Monitoring the environmental changes and correcting the data accordingly.
4. Artificially controlling the transducer environment to minimize these errors.

Many of the individual errors may be random in nature. The total error will usually be less than the algebraic sum of these individual errors. The use of an error band tends to simulate actual conditions because only the total accuracy deviation is considered, rather than the individual factors contributing to it. Some individual errors are predictable and can be calibrated out. When the system is calibrated, the calibration data can be used to correct the record data.

Environmental errors can be corrected by data reduction methods if the environmental effects are recorded simultaneously with the data. Then the data can be corrected by using the environmental characteristics of the transducer.

These techniques may provide a significant increase in system accuracy. They are useful when individual errors are specified and when the error band specification is not used.

In this book we seek to help the automation systems designer in choosing a transducer or sensor by providing relevant information for making the choice,

18 TRANSDUCERS FOR AUTOMATION

including definitions of specifications and features of selection and evaluation. We now move on to consider what must be done to make the total system perform as expected.

A key factor in choosing the right transducer is to completely define the design objectives. Consider all known objectives and try to anticipate the unknowns. Include such factors as signal and noise levels, desired accuracy, throughput rate, characteristics of the interfaces, environmental conditions, and size and system budgetary limitations that may force performance compromises or a different system approach.

Some general considerations for the measurement system include

1. An accurate description of input and output signal range; source or load impedances, type of digital code; logic level and logic polarity.
2. Data throughput rate.
3. Interface specifications.
4. System error budget allowed for each functional block.
5. Environmental conditions.
6. Supply voltage, recalibration interval, and other operating requirements.
7. Special environmental conditions; rf fields, shock, and vibration.

Low-level signals may become obscured by noise, rfi, ground loops, power-line pickup, and transients coupled into signal lines from machinery. Separating the signals from these effects becomes a matter for concern.

Most automation systems can be separated into two basic categories: those suited to favorable environments like processing laboratories and those required for hostile environments such as factories and petrochemical installations. The latter group includes those industrial processing systems where temperature information may be developed by sensors on tanks, boilers, vats, or pipelines that may be spread over acres of facilities. The data may be sent to a central processor to provide real-time process control. The digital control of steel mills, automated chemical production, and machine tools are characterized by this environment. The vulnerability of the data signals requires the use of isolation and other techniques.

In the laboratory type of environment, as in test systems for gathering information on gas samples, the designer is concerned with making sensitive measurements under favorable conditions rather than with the problem of protecting the integrity of collected data.

Systems in the more hostile environments can require

1. Data conversion at an early stage.
2. Shielding.
3. Common-mode noise reduction.

4. Components for wide temperatures.
5. Preprocessing of the digital data to test their reliability.
6. Redundant circuits for critical measurements.

Laboratory environment systems will have narrower temperature ranges and less ambient noise, but the higher accuracies will require more sensitive devices and an effort may be necessary to preserve the required signal/noise ratios.

The configuration and components in the data acquisition system require the consideration of a number of factors including the resolution and accuracy of the data in the final form, the number of analog sensors to be monitored, the sampling rate desired, the signal conditioning required due to the environment and accuracy needs, and the cost allowance.

Some of the choices for a basic data acquisition configuration include a single-channel system with direct conversion; preamplification and then direct conversion; a sample hold and conversion; preamplification, sample hold, and conversion; preamplification, signal conditioning, and direct conversion; or preamplification, signal conditioning, sample hold, and conversion.

A multichannel system could use the following techniques: multiplexing the outputs of single-channel converters, multiplexing the outputs of sample holds, multiplexing the inputs of sample holds, multiplexing low-level data, or more than one tier of multiplexers.

Signal conditioning can include ratiometric conversion or one of the wide dynamic range techniques such as high-resolution conversion, range biasing, automatic gain control, or logarithmic compression. Noise reduction may require analog filtering, integrating converters, or digital data processing.

In order to consider these techniques we first examine some of the components used in data acquisition. When several analog transducer signals are to be processed by the same microcomputer or communications channel, a multiplexer is used to channel the input signals into the A/D converter.

Multiplexers may also be used in reverse. If a converter must distribute analog information to many different channels, the multiplexer can be fed by a D/A converter which continually refreshes the output channels with new information.

In many automation systems, the analog signal may change during the time that the A/D converter takes to digitize an input signal. The changes in this signal level during the conversion process can result in errors since the conversion period can be completed some time after the conversion command and the final value never represents the data at the instant when the conversion command is transmitted.

Sample hold circuits can be used to make an acquisition of the varying analog signal and hold this signal for the duration of the conversion process. Sample hold circuits are often used in multichannel distribution systems, where they allow each channel to receive and hold the signal level.

20 TRANSDUCERS FOR AUTOMATION

In order to put the data in digital form as rapidly and as accurately as possible, we can use an analog-to-digital converter, which might be a shaft encoder, a small module with digital outputs, or a high-resolution, high-speed panel instrument. These devices, which range from IC chips to rack-mounted instruments, convert analog input data, usually voltage, into an equivalent digital form.

CONVERTER APPLICATION

Converters for automation system applications usually require external commands to convert or hold. For low-frequency signals, the converter may be an integrating type, which is inherently a low-pass filter. It is capable of averaging-out high-frequency noise and nulling those frequencies harmonically related to the integrating period. The integrating period can be made equal to the period of the 60-Hz line frequency, since a major portion of interference occurs at this frequency and its harmonics.

If the converter must respond to individual samples of input, the maximum rate of change of the average input, the full-scale voltage, and the conversion time (T_c) have the following relationship for binary conversion:

$$\left.\frac{dV}{dt}\right|_{max} = 2^{-n}\frac{V_{FS}}{T_c}.$$

If individual samples are not important but large numbers of samples are to be used (this is essentially a stationary process) the only requirement is that the signal be sampled at least twice each cycle for the highest frequency of interest.

The dual-slope integrating A/D converter spends about 30% of its sampling period performing an integration and the remainder of the time counting out the average value over the integrating period and resetting for the next sample. The dual-slope type will always read the average value, which results in a sample of the input waveform over the integrating time period. The integrating A/D converter is slow but useful for measurements of temperature and other slowly varying voltages, especially in the presence of noise.

The successive-approximation device can provide high resolution and high speed at a reasonable cost. If T_c, using a successive-approximation converter, is 10 μs, the maximum allowable dV/dt is 500 V/s. The successive-approximation converter will, at high rates of change, generate linearity errors since it cannot tolerate any changes during weighting. The converted value will be between the values at the beginning and the end of conversion, while the time uncertainty can be close to the conversion interval. If the signal is slow, noise with rates of change that are large can cause errors that may not be averaged

INTRODUCTION 21

during the conversion. An external sample hold can be used to improve this condition.

The process of selecting an A/D converter system is similar to that required for D/A converters. Some of the following considerations are analogous to those for D/A converters, while others are unique:

1. The analog input range and resolution of the transducer signal.
2. The requirements for converter linearity error, relative accuracy, stability and calibration.
3. The time allowed for a complete conversion.
4. The various sources of error that must be minimized as the ambient temperature changes such as missed codes.
5. The characteristics of the system power supply.
6. The type of reference required.
7. The input signal noise level. Will the input be sampled or filtered? Is it rapidly or slowly varying?

Then consider the conversion circuits acceptable for the application. Integrating types are best for noisy input signals at slow rates, while successive approximation is best for sampled or filtered inputs at rates to 1 MHz. Counter comparator types are cheap but tend to be slow and susceptible to noise.

When a system is assembled in which one A/D converter is time shared among input channels by a multiplexer and sample hold, their contribution to system errors must be considered.

In addition to these considerations, there are specific requirements to consider for the other parts of the system. For D/A converters:

1. Consider the resolution; the number of bits of the incoming data word must be converted as well as the analog accuracy and linearity that are required.
2. The digital code and logic level must also be selected.
3. The output signal may be a current or a voltage. Voltage output D/A converters are more convenient to use. Current output D/A converters are used in applications where high speed is more important than a voltage output, such as A/D circuits with comparators.
4. The reference may be fixed or variable, internal or external, or multiplying with a number of quadrants.
5. The speed requirements must be considered including the shortest time between data changes and after a change in the input, and the time the system can wait for the output signal to settle after a full-scale change.
6. Consider the switching transients and how they can be filtered.
7. Consider the temperature range including internal temperature rises and

the range over which the converter can perform within specifications without readjustments.
8. The power supply must be sensitive enough to hold errors to acceptable levels.

MULTIPLEXING

Most multiplexers use a MOSFET switch where the insulated gate is driven to a fixed potential in the on condition. The on resistance will vary with the level of the applied signal. A typical P-channel device may have $R_{on} = 100$ ohms at $+10$ V, and $R_{on} = 500$ ohms at -10 V. FET devices have a leakage from drain to source in the off state and a leakage from gate or substrate to drain and source in both the on and the off states. The gate leakage in MOS devices is small compared to other sources of leakages. When the device has a Zener-diode-protected gate, an additional leakage path exists between the gate and source.

Enhancement-mode MOSFETs have the advantage that the switch turns off when power is removed from the MUX. Junction FET multiplexers always turn on with the power off.

A more recent development, the complementary MOS (CMOS) switch has the advantage of being able to multiplex voltages up to and including the supply voltages. A ± 10-V signal can be handled with a ± 10-V supply.

There are a number of tradeoff considerations for the automation systems designer. Analog multiplexing has been the favored technique for achieving lowest system cost. The decreasing cost of A/D converters and the availability of low-cost, digital integrated circuits specifically designed for multiplexing provides an alternative with advantages for some applications. A decision on the technique to use for a given system will hinge on tradeoffs between the following factors:

1. *Resolution.* The cost of A/D converters rises steeply as the resolution increases, because of the cost of precision elements. At the 8-bit level, the per channel cost of an analog multiplexer may be a considerable proportion of the cost of a converter. At resolutions above 12 bits, the reverse is true and analog multiplexing tends to be more cost effective.
2. *Number of channels.* This controls the size of the multiplexer required and the amount of wiring and interconnections. Digital multiplexing onto a common data bus reduces wiring to a minimum in many cases. Analog multiplexing is suited for up to 256 channels; beyond this number, the technique is unwieldy and analog errors become difficult to minimize. Analog and digital multiplexing are often combined in very large systems.
3. *Speed of measurement, or throughput.* High-speed A/D converters can

add a considerable cost to the system. If analog multiplexing demands a high-speed converter to achieve the desired sample rate, a slower converter for each channel with digital multiplexing can be less costly.
4. *Signal level and conditioning.* Wide dynamic ranges between channels can be difficult with analog multiplexing. Signals less than 1 V generally require differential low-level analog multiplexing, which is expensive, with programmable gain amplifiers after the multiplexing operation. An alternative is fixed-gain converters on each channel, with signal conditioning designed for the channel requirement.

In digital systems a variety of devices may be used to drive the bus, from open-collector and tristate TTL gates to line drivers and optoelectronic isolators. Channel selection decoders can be built from 1 to 16 decoders to the required size. This technique also allows additional reliability in that failure of one A/D does not affect the other channels. An important requirement is that the multiplexer operate without introducing unacceptable errors at the sample rate speed. For a digital multiplexing system, one can determine the speed from propagation delays and the time required to charge the bus capacitance.

Analog multiplexers tend to be more difficult to characterize. Their speed is a function not only of internal parameters but also of external parameters such as channel source impedance, stray capacitance, number of channels, and circuit layout. The user must be aware of the limiting parameters in the system to judge their effect on performance.

The nonideal transmission and open-circuit characteristics of analog multiplexers can introduce static and dynamic errors into the signal path. These errors include leakage through switches, coupling of control signals into the analog paths, and interactions with sources. The circuit layout can compound these effects. Their relevant multiplexer considerations also include

1. The number and type of input channels needed: single-ended or differential, high or low level, and their dynamic range.
2. The type of hierarchy used for a large amount of channels: the addressing scheme.
3. The allowable crosstalk error between channels and the frequencies involved.
4. The channel switching rate: fixed or flexible, continuous or interruptible.
5. The settling time when switching from one channel to another, the maximum switching rate.
6. The errors due to leakage through source resistances.
7. The transfer errors due to the voltage divider formed by the on resistance of a multiplexer and the input resistance of a sample hold.
8. Source damage when the power is off: MOSFET multiplexer switches

24 TRANSDUCERS FOR AUTOMATION

open when power is removed. JFET multiplexer switches can conduct when power is removed. Thus, it is possible to interconnect and damage active signal sources.

The relevant sample hold considerations include

1. The input signal range.
2. The slewing rate of the signal, the multiplexer's channel switching rate, and the sample hold acquisition time.
3. The accuracy, gain, linearity, and offset errors.
4. The aperture delay and jitter. (The delay component of the aperture time may be correctible, since the switching can be advanced in time to compensate for this. The uncertainty or jitter cannot be compensated. In systems that use a constant sampling rate with data that are not correlated to the sampling rate the aperture delay is not important, but the jitter may modulate the sampling rate.)
5. The amount of droop allowable in the hold condition.
6. The offset errors due to the sample hold's input bias current through the multiplex switch and sources.
7. The effects of aging, temperature, and power supply variations.

It is essential to have an understanding of what the manufacturer means by the specifications. It should not be assumed that manufacturers mean the same thing when they publish the same numbers defining a parameter. Product information must be interpreted in terms meaningful to the user's requirements, which requires a knowledge of how the terms are defined. The specifications may not mean what the user thinks they mean and it is important to consider their implications.

In systems where the channels may be diverse rather than identical, the multiplexer could be switching sequentially or in a random selection mode. In some cases, manual operation may be desired for checkout purposes. With the random access mode, it is desirable that those channels with more intelligence be accessed more frequently. In addition to sharing the converter and the sample hold, the instrumentation amplifiers can be conserved.

The successful transmission and multiplexing of low-level data requires a different approach. Low-level multiplexing often uses programmable gain amplifiers, or automatic range switching preamps, which allows the use of converters having medium resolutions with range switching to obtain additional significant bits. For example, a 12-bit converter, and 32 steps of adjustable gain, could provide 17-bit resolution, if the resolution is actually present in the signal and the system can operate on it without degradation.

Low-level multiplexers are always differential or two wire, so the converter sees only the difference in the errors of two identical channels. Leakage, gain, and crosstalk effects are greatly reduced, provided that matching is maintained. The magnitudes of most settling errors are lowered although their duration remains the same.

NOISE AND INTERFERENCE

When multiplexers must be operated under conditions causing high common-mode interference, two-wire differential and guarded or flying capacitor multiplexers can be used. If considerable normal-mode interference also exists, further steps may be required such as filtering, digital averaging, and the use of integrating converters and digital multiplexing.

The use of low-pass filters in the channel inputs of the multiplexer is one method of reducing the normal-mode interference. Filter characteristics can be tailored for the channel. Filters may increase the settling time, but this effect is usually small. If the filter is placed after the multiplexer, then each channel will have to charge the filter, greatly increasing the settling time. In differential systems, the filters should have balanced impedance in both inputs or be connected differentially to reduce common-mode effects.

When passive filtering of each channel is not practical, an integrating A/D converter can provide high normal-mode rejection. This rejection is obtained with a conversion time that is usually shorter than the settling time of a filter required to provide the same rejection.

Rejections of normal-mode interference to 70 dB can be obtained with an integrating converter. Many integrating converters are designed for floating guarded inputs.

In automation systems where processing time and memory are available and if the converter can track the variations in input signal produced by interference, software can be used to reduce the effects of interference.

Multiple samples can be taken for each channel and the results summed and averaged. The signal noise ratio will improve as the square root of the number of samples, provided that the sampling and interference frequencies are not correlated.

By digitizing the sensor signals at their source, one can perform logical operations on the digitized data before they are fed into the system computer. Processing tends to be more streamlined and test problems are minimized.

The use of digital signals allows a considerable immunity to line frequency pickup and ground loop interference to be achieved. The digital signals may be transformer or optically coupled for complete electrical isolation. Also, the use of low-impedance digital driver and receiving circuits can drastically reduce the vulnerability to noise.

Preprocessing circuits or microcomputers can access data from slow thermocouple sensors less frequently, while reading data from faster, more critical sources at greater speeds. The digital subsystem can make its own decision as to when particular data should be fed into the main computer. Certain signal sources remain constant or are within a narrow range for long periods. They may then change rapidly later during the physical process and it may be useful to ignore the data until the changes occur.

Both flexibility and versatility can be gained by transferring the interface process from analog multiplexing to digital multiplexing. Decision circuits can exercise judgement on when and what data to feed the main computer, and this in general tends to improve the overall interface.

In some automation applications when the data are being transmitted over an appreciable distance the transmission channel may become crowded, and same sort of data compression is essential.

In the design of an automation system module, care should be used to separate the analog and digital signal lines. This should also be followed in the layout of the area in which the system is to be installed. Digital lines should not run parallel in close proximity with runs of analog signal lines. When these lines cross, they should do so at right angles. Care must be taken with low-level high-gain points, such as the comparator input in A/D converters and the summing junction of the output amplifier in D/A converters. Runs to these points should be short and not create loops. Ground guard runs can be used to reduce most interference.

Most data acquisition components have a number of ground terminals, which are not connected together in the module. These grounds may be referred to as the logic power return, analog common, analog power return, analog signal ground, or analog sense. These grounds should be tied together at one point, usually at the system power supply ground. A single solid ground is most desirable. However, current flows through the ground wires and tracks of the circuit cards, and since these paths will have some resistance and inductance, several hundred millivolts may be generated between the system ground point and the ground terminal of the module.

The connections between the system ground point and the ground terminals should be as short as possible and should have the lowest possible impedance. The module's supply terminals should be capacitively decoupled as close to the module as possible. A capacitor with a good frequency response should be used. A 15-μF solid tantalum capacitor is usually recommended. The analog supplies are bypassed to the analog power return terminal and the logic power terminal is bypassed to the logic power return. If gain and offset adjustments are required, the potentiometers should be mounted with short leads in a position that will be accessible.

Separate returns can be used to minimize the current flow from sensitive points to the system ground point. In this way, supply currents will not flow in the same return path with analog signals and logic return currents are not summed with other return currents.

Electronic devices should not be located near a transformer or fan motor. Using shielding to protect against interference is expensive and not always successful. D/A converters should always be located near their loads. This may require longer cable runs for the digital signals; however, the overall reduction in noise can justify the expense. A/D converters should be located as near the signal source as possible. One can also use a differential amplifier to receive the signal at the end of a long run before it enters the A/D converter. Unshielded analog should never be run near either digital or power lines.

To reduce common-mode errors, a differential amplifier may be used to eliminate ground potential differences as shown in Figure 1-5. The signal source may be a remote transducer with the differential amplifier located at the A/D converter. The common-mode signal is the potential difference between the ground signal at the converter and the ground signal at the transducer, plus any common-mode noise produced at the transducer and voltages developed by the unbalanced impedances of the two lines.

When the transducer signal source is the output of a D/A converter, the differential amplifier is located near the load. The common-mode signal will be a function of the differences in ground potential at the two locations. The amount of dc common-mode offset that is rejected will depend on the rejection ratio of the differential amplifier. Currents flowing through the signal source leads may cause offsets if either the currents or the source impedance are unbalanced.

At the higher frequencies, unbalanced conditions, series resistance, shunt capacitance, and the amplifier's internal imbalances reduce the common-mode rejection, producing a quadrature normal-mode signal. This error can be reduced by the use of a shield, as shown in Figure 1-6. In this circuit, no part of the common-mode signal appears across the line capacitors (C_L) since the shield is driven by the common-mode source. The shield also provides some electrostatic shielding to limit coupling to other lines in close proximity.

It is important that the shield be connected only at one point to the common-mode source signal and that the shield be continuous, through all connectors. The shield is carrying the common-mode signal, so it should be insulated to prevent it from shorting to other shields or ground. A return path must exist for the leakage currents of the differential amplifier unless it has transformer or optically coupled inputs.

In this chapter we have discussed the system aspects of applying transducers. Considering the different types of transducers on the market and the complex manner in which transducer specifications may relate to a system application,

28 TRANSDUCERS FOR AUTOMATION

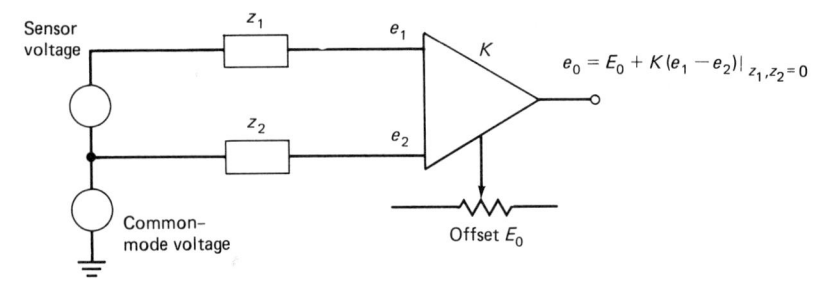

Figure 1-5. Using a differential amplifier to reduce common-mode errors.

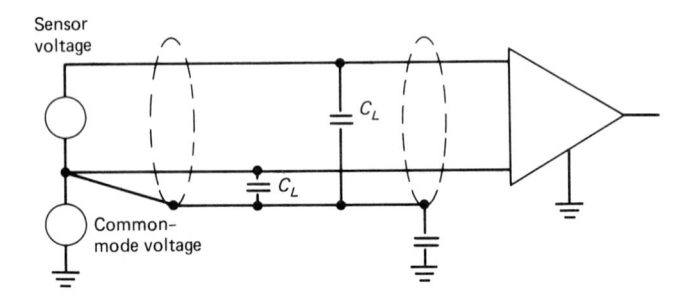

$$e_0 = E_0 + K(e_1 - e_2)|_{z_1, z_2 = 0}$$

Figure 1-6. Using a shield in higher-frequency systems.

selecting the best transducer for an automation application is not always a simple task. To make the most appropriate choice, we must consider a number of issues: the objectives of the automation conversion process and how they relate to the transducer's specifications, how the system must be configured to meet the performance requirements, and how the other system components limit and degrade transducer performance and affect tradeoffs in the system error budget.

A relaxation of the class of errors due to environment-related specifications can be achieved by allotting one multiplexer channel to carry a ground-level signal, and another to carry a precision reference voltage level that is close to full scale. Data from these channels are used by a microcomputer to correct gain and offset variations common to all the channels. These errors may be generated in the sample hold, A/D converter, or wiring.

Upon selecting the appropriate transducer the user should be aware that the analysis involved usually is not by itself sufficient to ensure proper system performance. The automation system designer must consider the physical interconnections, grounding, power supplies, protection circuitry, and other details that constitute good engineering practice. To evaluate the performance tradeoffs, an error budget can be used. Three classes of errors can be considered, those due to the nonideal nature of the devices, the physical interconnections, and the interaction of devices. The first type of error is determined from the specifications of the devices that we discussed in the first part of this chapter.

EXERCISES

1. List the properties of some typical transducer specifications.
2. Describe some techniques used for the isolation of noise in automation systems.
3. Discuss some methods of reducing the total error in a measurement system.
4. In general, the objective of transducer selection is to choose the least expensive device that will meet the physical, electrical, and environmental requirements for the application. What detailed criteria will determine the best choice in those applications where the desired performance requirements can be met by several transducers?
5. The difficulties in single-channel low-level data acquisition are compounded by the addition of low-level multiplexing of such channels. Discuss the use of shields and shield grounds in a low-level multiplexing system.
6. Discuss the tradeoffs between using digital multiplexing before transmission or remote A/D conversion and serial transmission.
7. Discuss the use of sample holds in data acquisition. In what ways may they be used to improve performance?
8. Consider an error budget analysis. Discuss its importance in analyzing the error problem and applying error reduction to the most important sources of error in an automation system.
9. What circuit resistances control noise and interference in a data acquisition system? How can these be minimized?

2
TEMPERATURE MEASUREMENT

INTRODUCTION

Temperature is a particular condition of a substance. The classical definition depicts heat as a form of energy associated with the activity of the molecules of the substance. The particles are assumed to be in a state of continuous motion which is sensed as heat. The temperature is a measure of this heat.

For a particular measurement application, there may be one type of sensor that meets the performance and economic requirements better than all others.

In the design and development of an automation system, an increased accuracy of measurements can result in more successful and economical system operation. When computer control or remote readouts are used, the long lines required can often lead to difficulty in using the low output of some temperature sensors without amplification. This is where resistance thermometers can be used, since they have a relatively high output and good linearity.

RESISTANCE TEMPERATURE DETECTORS

Resistance temperature detectors (RTDs) utilize the temperature dependence of the resistance of a material to electric currents. The resistance of metals increases with increasing temperature, while most semiconductor materials decrease in resistance. A part of the total resistivity in metals is due to the impurities in the metal. This is the residual resistivity. It is lowest for the pure metals.

The residual resistivity can change if the detector is used at too high a temperature, or if the wire is contaminated by the environment or materials in contact with the wire. The change is relatively independent of temperature. It becomes apparent when the resistance difference $(R_T - R_0)$ between the two temperatures remains constant while the ratio (R_T/R_0) decreases. These changes are irreversible.

Another part of the metal's resistivity is due to deformation, which depends on the physical state of the metal. In a well-annealed metal with a low resistivity, as the crystal structure is stressed, the resistivity increases.

The resistance can be increased by mechanical shock or vibration, thermal shock, or nuclear radiation at low temperatures. A difference in thermal expansion coefficients between the wire and its supporting structure (see Fig. 2-1) can cause stress on the wire. The RTD then acts as a strain gauge as well as a

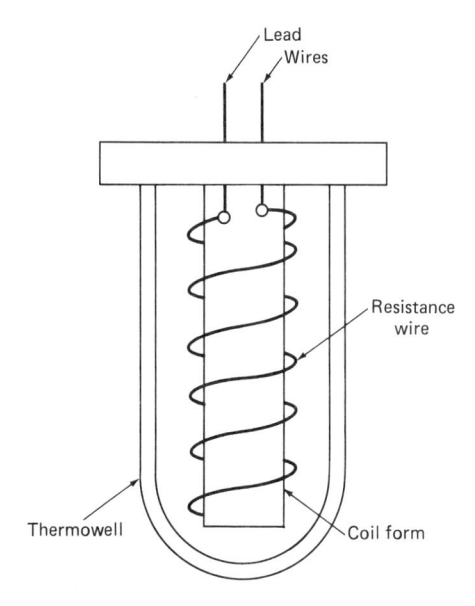

Figure 2-1. Typical RTD construction.

temperature transducer. This effect is reversible and the original resistance can be restored by reannealing the resistance element. Additional materials such as cements used should not introduce strains over the operating range.

In very high-temperature applications changes in the wire dimensions due to evaporation can occur. In this case, R_T/R_0 remains constant but $R_T - R_0$ increases.

These effects must be considered in both the design and use of a RTD. The elements must be designed such that the wire remains annealed and strain free for the operating temperature range. By minimizing the strain, the R vs. T characteristic remains similar to that of the wire alone; the repeatability can be as sensitive as 0.1°C for industrial sensors and 0.001°C for platinum wire standards. Strain-free designs are manufactured by matching the expansion characteristics and by allowing the wire to expand or contract freely as the temperature changes. The materials that are used must be compatible with both the environment and the sensing element. This includes thermal shocks and mechanical vibration. In some applications, these devices will tend to age in use until stress are equalized.

Copper is an inexpensive material for resistance temperature sensing and is one of the most linear metals for a wide temperature range. It has a tendency to oxidize at moderate temperatures along with exhibiting poor stability and reproducibility compared to platinum. The low resistivity of copper is also a disadvantage.

Nickel has been widely used as an RTD element over the temperature range from $-100\,°C$ to $300\,°C$, mainly because of its low cost and high temperature coefficient. Above $300\,°C$, the R/T relation for nickel changes greatly. Nickel is also susceptible to contamination by certain materials such as sulfur and phosphorus. It has an R/T relation that is neither as well known nor as reproducible as that of platinum.

The resistance temperature characteristics of tungsten is also not as well known as that of platinum. Fully annealed pure tungsten is upredictable owing to the brittleness caused by recrystallization. Tungsten sensors tend to be less stable than platinum sensors. Tungsten does have a good resistance to high nuclear radiation levels, as does platinum. The mechanical strength of tungsten allows one to draw fine wires and the resulting sensing units can have high resistance values.

PLATINUM RESISTANCE SENSORS

Platinum resistance thermometers provide high accuracy from the triple point of hydrogen (13.81 K) to the freezing point of antimony ($630.74\,°C$). The resistance/temperature relationship of platinum is well known and tends to be reproducible and linear over this temperature range. Platinum is chemically inert and not easily contaminated since it does not readily oxidize. It can be used up to $1500\,°C$. Platinum RTDs tend to be more expensive than other resistance temperature sensors, although some industrial types are competitive.

Thin-film platinum temperature sensors use a deposit of a thin film of platinum on an insulating substrate. This technique allows a high resistance in a small sensing package with a relatively fast response time. The fabrication techniques are similar to those used to manufacture integrated circuits. Platinum wire RTD elements tend to be delicate and repeated thermal cycling can cause aging. The thin-film deposition technique used with laser trimming of the resistance can result in a small, rugged sensor at an attractive cost. Any differences in applying these high-purity films to the substrate causes the resistance/temperature characteristic to behave differently from that of pure, annealed, strain-free platinum. The temperature coefficient at room temperature is usually about 30 to 80% that of pure platinum. The residual resistance can be much higher. The variation of the resistance temperature characteristic between devices can be a potential problem is some automation applications. Thin-film platinum detectors can exhibit an accuracy and repeatability of less than $0.01\,°C$. The small sensing tip allows a more rapid response time than thermocouples. A glass–capsulated platinum layer is usually deposited on a ceramic substrate with dimensions of $10 \times 3 \times 1$ mm. The rapid response times result from a large surface area to volume ratio as well as the use of a thin ceramic substrate which increases the thermal conductivity.

THERMISTORS

Thermistors are mixtures of semiconductor materials with a resistance that varies rapidly with temperature. These materials are usually sintered mixtures of sulfides, selenides, and oxides of metals such as nickel, manganese, cobalt, copper, and iron. They are formed into small glass-enclosed beads, disks, or rods as shown in Fig. 2-2.

High resistivity and high negative temperature coefficient are typical of these sensors, although the temperature coefficient can be positive as well. The sensors with a positive temperature coefficient are used at high temperatures and as current limiters.

The resistance versus temperature characteristic is nonlinear, and it has been difficult in manufacturing these devices to maintain narrow resistance tolerances, although interchangeable devices are now possible using advanced manufacturing processes.

Thermistors are relatively inexpensive and available in small sizes with high resistance values. If a burn-in calibration procedure is used, the stability is as good as the best wire-wound resistance sensors. The nonlinear T vs. R relationship of thermistors requires numerous calibration points and this can be a major cost factor in a thermistor application. It has been difficult to produce thermistors with good interchangeability, but selected thermistors or matched pairs in series are now available which are reasonably interchangeable.

A single thermistor is normally unsuited for wide temperature spans because of its limited linear resistance characteristic, so several thermistors are used to cover a wide temperature span.

Composite thermistor resistor assemblies can be used to provide linear response curves. External linearization is not required. Composite thermistor sensors can be obtained in standard ranges: -5 to $45°C$, -30 to $50°C$, and 0 to $100°C$. These networks can have a higher sensitivity than thermocouples.

Components are also available as composite networks of thermistors and resistors packaged as a single sensor. These chip thermistor networks can be bonded directly to hybrid substrates for temperature sensing.

Another technique uses resistor pastes which are screened onto alumina ceramics. The pastes, which may have either positive or negative temperature co-

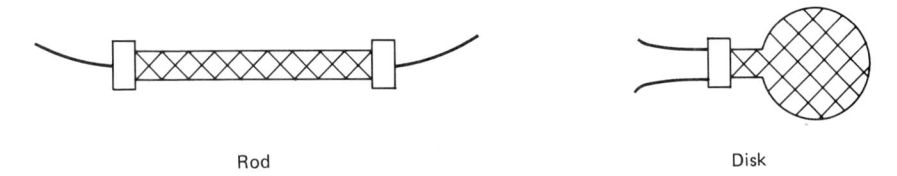

Rod   Disk

Figure 2-2. Thermistor constructions.

efficients, are then fired to produce temperature-sensitive resistor elements with relatively good linear responses. High- and low-resistivity paste formulas are available, with resistivity ranges from 100 ohms/☐ to over 1 megohm/☐.

Other semiconductor sensors include germanium and silicon crystals and carbon resistors. Germanium crystals have been used for temperature measurements in the 1 to 35 K region. A single crystal of germanium that has been doped with controlled impurities is used. Repeatability can approach 0.001 °C near the helium boiling point and 0.05 °C near the hydrogen boiling point. The resistance versus temperature characteristic has a nonlinearity similar to that of thermistors. Calibration interchangeability is difficult to obtain.

Carbon resistors have been used for temperature sensing in the region below 60 K, where they behave similarly to silicon semiconductors. High resistivity and negative temperature coefficient of resistance are characteristic of this type of sensor. Disadvantages include sensitivity to pressure variations in pressurized applications, long-term resistance drift, short-term drift when cycling between low and room temperatures, and a relatively long response time.

Crystals of silicon that have been properly doped have the characteristics required for resistive temperature sensors. The temperature coefficient of resistance is positive above −50°C and the resistance versus temperature characteristics are almost linear, especially over the most usable range of −50°C to 250°C. Below −50°C the temperature coefficient of resistivity is negative and the slope of the resistance versus temperature characteristic increases quickly.

One type of silicon crystal sensor is made of heavily doped p-type silicon, which provides a positive temperature coefficient of 0.7%/°C. Unlike a typical thermistor, the linearity can be within ±0.5°C without requiring a linearization network. Over a range of zero to 100°C, the linearity can be within ±0.025°C. Resistance values start at 10 ohms and progress through 40 values up to 10 kohms.

EXCITATION METHODS

Since resistance sensors do not produce voltages, they must be energized from external power supplies. Excitation is complicated by the fact that the resistance cannot assume an arbitrary zero value at any temperature since it changes value from R_{T1} to R_{T2}. To obtain a zero-based ouptut signal at a reference temperature the sensor is normally used as an arm in a resistance bridge network.

The Wheatstone bridge as shown in Figure 2-3a can only be used when the sensor's resistance value is high enough to mask the resistance due to the lead wires joining the bridge to the sensing element. This is true in the case of thermistors with their higher resistance.

Cable runs to connect the sensor to the bridge in an automation system may vary from a few feet to several hundred feet in length. If the bridge is installed in the same environment and we may not gain any advantage.

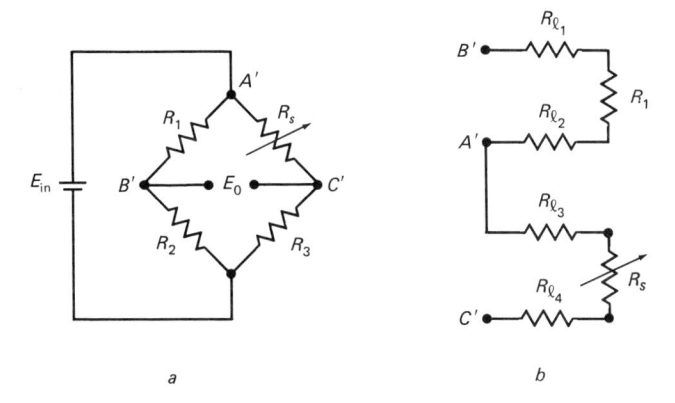

Figure 2-3. a. Wheatstone bridge. b. Compensation loop.

Temperature-sensitive wire must be used to connect a temperature-sensitive resistor so we must measure both of these sources. A number of methods can be used for bridge excitation.

Either ac or dc bridge excitation can be used depending on how the output signal is to be used in the system. Alternating-current excitation is used in some systems because of the type of recording and control instrumentation used. One advantage is the elimination of thermal emfs in the sensor leads. However, reactive effects such as the phase shifts from the cable capacitance and noise pickup and other interference due to cable coupling can prove to be troublesome. Thermal emfs are normally insignificant.

Direct-current excitation allows unwanted noise to be filtered out. The accuracy and stability of the dc power supply used directly affect the accuracy and stability of the bridge output signal. Power supplies with stabilities to 0.01% can be required in some high-performance applications, although a stability of 0.1% is usually adequate for most automation applications.

A compensation loop for reducing the effects of lead wires is shown in Figure 2-3b. Another technique is to use three leads. In this circuit, one lead is connected in series with the power supply where any resistance changes have negligible effects on the voltage. The two remaining leads are placed in series with an opposite leg of the bridge where their effects tend to cancel.

TRANSISTORS AND INTEGRATED CIRCUITS

Transistors can also be used as temperature sensors. The most common technique is to use the change in emitter-to-base voltage. A properly characterized silicon transistor can produce a linear change in emitter-to-base voltage over the $-40°C$ to $150°C$ range. The correlation between the temperature and the transistor base-to-emitter voltage is approximately 2 mV/°C. Most sensors use a 400-mV total output change and accuracy is ±2% or 5°C. The thermal time constant for a typical sensor of this type is 3 s in a flowing liquid and 8 s in moving air. A stable constant-current source must normally be used.

IC temperature transducers may contain the complete compensation circuit required along with the sensor itself. These sensors may use laser trimming to produce a precalibrated temperature transducer for automation applications below 150°C. These sensors typically operate as a current source and provide an output of one microampere per degree Kelvin. The trimming of internal calibration film resistors during manufacture allows a ±0.5°C accuracy and linearity within ±0.3°C over the full range. The output impedance can be greater than 10 megohms, which allows the rejection of much of the hum, ripple, and noise. The high output impedance allows these sensors to be used in remote sensing systems with lines that may be many hundreds of feet long. These sensors can also be powered directly from 5-V logic, which allows simple interfacing to microprocessor systems.

Some IC sensors output the temperature changes as a frequency change. A voltage-to-frequency converter is used to provide a digital pulse output. They may be trimmed for Celsius, Kelvin, or Fahrenheit measurements using external components. The pulse train is proportional to the substrate temperature of the IC.

THERMOCOUPLES

The basic thermocouple circuit is shown in Figure 2-4. It consists of a pair of wires of different metals joined or welded together at the sensing junction and terminated at their other end by a reference junction. This junction is maintained at a known temperature called the reference temperature. The load resistance of the signal-conditioning or readout equipment completes the circuit.

When a temperature difference exists between the sensing and the reference junctions, a voltage is produced which causes a current to flow through the

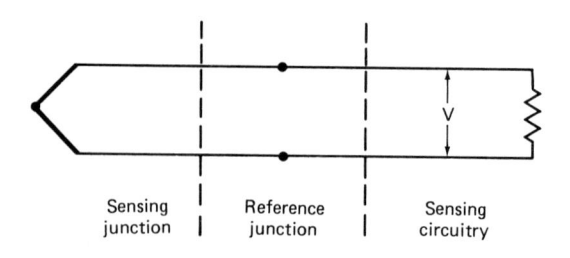

Figure 2-4. Basic thermocouple circuit.

circuit. This thermoelectric effect is due to the contact potentials at the junctions. It was originally discovered by Seebeck and is known as the Seebeck effect.

The wires between the sensing junction and the reference junction must be made of the same material. The connecting leads from the reference junction to the load resistance are usually copper. They are required to be copper whenever the associated wiring in the signal-conditioning or output circuits is copper to avoid any error-producing junction potentials due to the different metals.

In addition to the Seebeck effect there are the Peltier and Thomson effects. As current flows across a junction of two dissimilar conductors, heat unrelated to I^2R heating is absorbed or liberated as a function of the direction and magnitude of current. This is the Peltier effect. When a current flows through a wire along which a temperature gradient exists, heat unrelated to I^2R heating is absorbed or liberated from the wire. This is the Thomson effect.

The magnitude of the thermoelectric potential produced depends on the wire materials selected and on the temperature difference between the two junctions. The most common thermocouple materials are Chromel-Alumel, iron-Constantan, copper-Constantan, and Chromel-Constantan. Tables showing the thermal emf versus temperature for most of the standard materials have been developed by the National Bureau of Standards, the Instrument Society of America, and many of the thermocouple manufacturers. Most of these are based on a reference temperature of 0°C. The basic laboratory method uses an ice bath for the reference temperature, but this has limitations whenever measurements must be made frequently or continuously.

Techniques which can be used in place of the laboratory ice bath include automatic ice-point references, automatic oven references at temperatures other than 32°F (usually 150°F), and electrical compensators. Most electrical compensators use an ambient temperature couple in a bridge circuit with ambient sensitive elements. A typical electrical compensator is shown in Figure 2-5. This is an economical method for single-channel measurement systems with accuracies of ±1°F.

The oven type of reference can be used for a number of thermocouple channels. The oven references, as well as the ice-point references, are cost effective whenever a number of reference couples are required.

If the leads from the thermocouples are brought to a switch box, the contact resistances of the switches can cause errors. Any nonconducting film on the contacts can be a problem since only a few millivolts are available to penetrate this film.

A number of precautions should be taken when installing thermocouples:

1. Always locate the thermocouple in an average-temperature zone.
2. The thermocouple should be completely immersed in the medium so a true temperature can be measured.

38 TRANSDUCERS FOR AUTOMATION

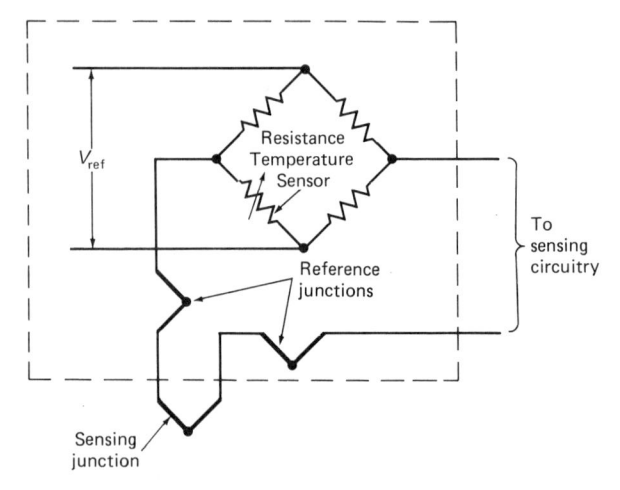

Figure 2-5. Thermocouple compensation bridge.

3. Do not locate the thermocouple near or in a direct flame path.
4. Always try to locate the thermocouple where the hot junction can be seen from an inspection port in high-temperature zones.
5. Avoid cable runs parallel to or closer than one foot to ac supply lines since any induced currents can cause errors.
6. All connections must be clean and tight in order to avoid errors due to contact resistance.

The use of a pocket or well will impede the transfer of heat to the sensing element and reduce the speed of response of the sensor. Agitation of the fluid around the sensing element tends to reduce this effect and improves the speed of response.

Another potential problem is heat transfer. When there are two bodies and one is hotter than the other, there will be a net transfer of energy from the hotter body to the colder one. When a number of bodies are all at the same temperature and enclosed in a space insulated to heat, each body is considered to radiate energy into the surrounding medium and continuously absorb energy at the same time. Since there must be equilibrium, the processes balance one another, and the temperature of each body remains constant. This is known as the Prevost theory.

If we have a chamber where the walls are at a lower temperature than the internal hot gases, a thermocouple placed in the hot gas stream will be exposed to the heat of the hot gases; since it is at a higher temperature than the walls it radiates more heat than it receives and may measure a lower value than the true temperature of the gases. This phenomenon must be considered when locating temperature-sensing devices in these situations.

A self-heating error can occur if excessive power is dissipated in either the standard or the unknown. This can cause an error by heating the resistive element.

A thermal lag error can exist if the standard and the unknown tend to respond to a changing temperature at different rates or if the reference bath is not stable with respect to time.

Stem conduction errors may occur as a result of heat transfer along the stem of the standard or the unknown. The thermocouple should always be immersed ten inches into a liquid. For sensing devices with a short stem, the stem conduction error can be found by measuring the temperature at a number of depths.

Errors can also exist as a result of the uncertainty of temperature in an unmonitored bath. In an ice bath made with tap water, the temperature can deviate from 0°C by as much as 0.5°C in some regions. The use of an ice bath after too much ice has melted will also cause errors. Automatic ice-point devices which maintain the thermocouple at the ice point, as opposed to the introduction of an equivalent voltage, tend to be the most accurate references. These are available in both single- and multiple-channel units with accuracies of ± 0.1°F. Their long-term stability is excellent since they operate with the reference thermocouples located within a hermetically sealed cell in which the ice water equilibrium is maintained automatically.

The newer thermocouples react more rapidly than the older units because of their smaller thermal mass. Their typical response is in milliseconds. Wires 0.0008 in. in diameter are used to form the junction. These are then inserted into a quartz insulator which is assembled into a 0.008-in.-o.d. metal sheath or probe. The small diameter junction of less than 0.002 in. provides the faster response.

Errors due to radiation and conduction are also reduced, but the traditional problems remain. The voltage output is small and nonlinear and absolute measurements require a known reference. Expensive cabling and connectors are required to avoid unwanted junctions.

The thermopile is a combination of several thermocouples of the same materials connected in series. The output of a thermopile is equal to the total output of the number of thermocouples in the assembly. All reference junctions must be at the same temperature as shown in Figure 2-6.

The two major temperature sensors have been resistance thermometers and thermocouples. In the final selection process an accuracy and cost comparison can be made for the total system including sensors, signal conditioning, and readout circuits. The advantages of resistance thermometers over thermocouples include the following.

1. A reference junction temperature or compensating device is not required.
2. A larger output voltage is obtained, typically increased by a factor of 500 or more.

40 TRANSDUCERS FOR AUTOMATION

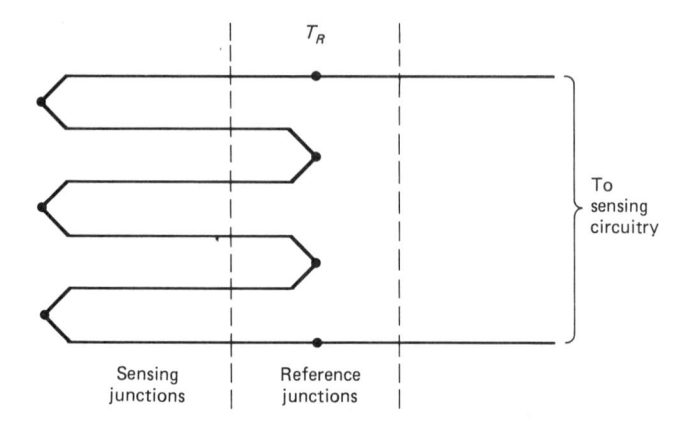

Figure 2-6. Thermopile circuit.

3. Recording, controlling, or signal conditioning equipment can be simpler, more accurate, and less expensive because of the greater output signal.
4. Electrical noise is not as great a problem with resistance sensors and longer lead wires can be used.
5. The sensitivity to small temperature changes is greater.
6. The output voltage per degree for resistance sensors can be selected by adjusting the excitation current and/or the bridge design.
7. The shape of the curve of output versus temperature can be modified by changing the resistance sensor bridge design.
8. In moderate temperature applications, the absolute accuracy of calibration as well as the stability of calibration for resistance elements can be higher by a factor of 10 to 100.

The major problem in the use of thermocouples is the errors due to the spurious and parasitic emfs in the leads. This effect is responsible for the difference in precision between thermocouples and resistance thermometers. Variations in the state or composition of a wire tend to produce Seebeck-generated emfs wherever a temperature gradient exists. This can become critical in applications where the temperature gradient on the leads is changing. Annealing reduces the causes of the effect but does not eliminate it. A periodic recalibration of the thermocouples is needed to offset the temperature distribution.

RADIATION TECHNIQUES

When the temperatures are too high to allow a thermocouple or other contacting temperature sensing element to be used, noncontact methods must be used. This technique is based on the fact that all hot bodies emit radiant energy with an intensity that is proportional to the absolute temperature of the emitting surface. Radiation pyrometry provides a way to measure the heat radiated by a hot object. It is possible to use this method for almost any hot object except for pure inert gases like helium.

In practice, radiation pyrometers are sensitive to a limited wavelength band of radiant energy. The operation of practical thermal radiation pyrometers is based on blackbody radiation concepts (Fig. 2-7). Some of the noncontacting instruments respond to a relatively wide band of wavelength and operate according to the Stefan-Boltzmann law. These are referred to as total-radiation pyrometers. Another group of instruments uses narrow bands of wavelength in the visible spectrum. Instruments of this type are known as optical pyrometers.

In an optical pyrometer as shown in Figure 2-8, the radiation source is viewed through a telescopic system consisting of a lens and eyepiece. Inside the system is a small lamp. The current through the lamp is controlled. An optical filter is placed between the eyepiece and lamp. As the operator looks through the eyepiece, the source is seen as a circle, square, or other easily recognizable shape. In the center of this shape is the image of the lamp filament. The control resistance is then adjusted until the brightness of the filament equals that of the radiation. The filament image will then appear to merge into the radiation image and present a uniform image to the operator.

The control knob may be calibrated directly in degrees, but in recent instru-

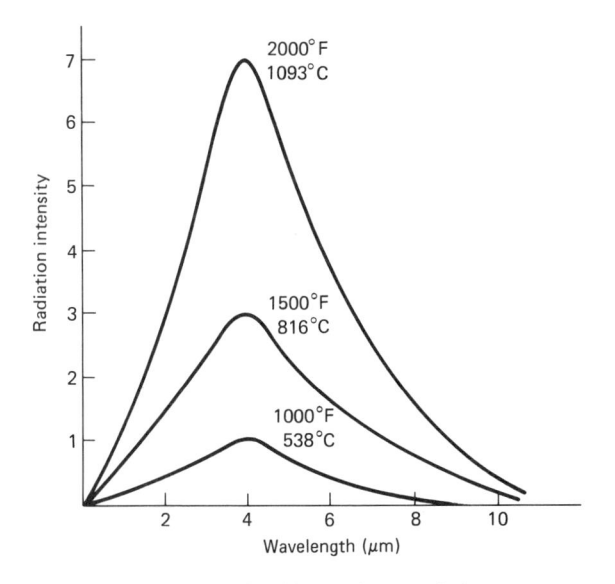

Figure 2-7. Blackbody characteristics.

42 TRANSDUCERS FOR AUTOMATION

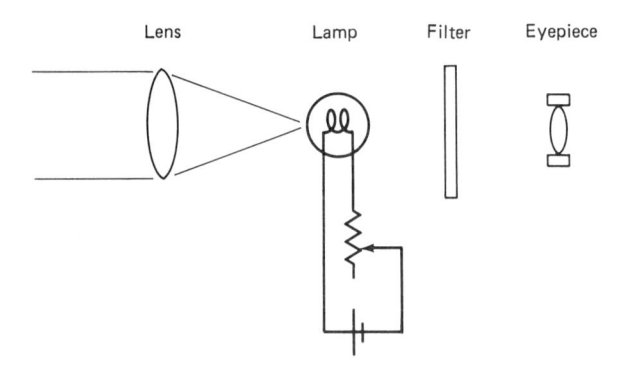

Figure 2-8. Optical pyrometer.

ments the operator control function has been replaced by electronic techniques which eliminate operator-to-operator differences.

A total-radiation instrument is shown in Figure 2-9. In this instrument radiation from the source is focused on a small disk of blackened platinum or a similar blackbody. In some units a concave mirror is used for focusing. Attached to the platinum disk are a number of thermocouples connected in series as a thermopile and connected to the sensing elements. Photomultiplier tubes and photodetectors are also used in some units. The radiation from the source heats the disk and generates an emf which is then converted to degrees.

In some total-radiation instruments the cold junction is very close to the hot junction. This placement of the hot and cold junctions ensures that both are almost equally affected by any ambient temperature changes. The cold junction is shielded, however, from radiation from the source. The use of fine thermocouple wire reduces conduction losses from the couple itself.

To prevent an excessive temperature rise of the pyrometer housing, the housing is enclosed in an insulated case. For cold junction compensation, a nickel

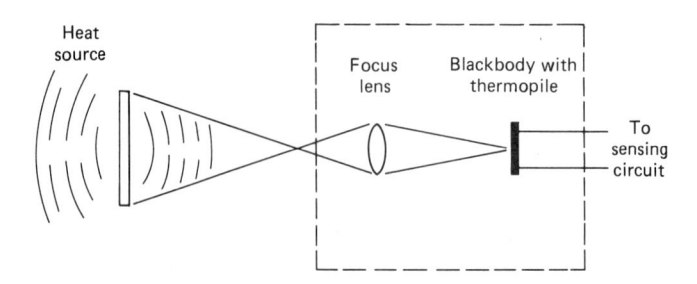

Figure 2-9. Total-radiation pyrometer.

resistance spool may be connected as a shunt across the thermocouple leads at the cold junction end.

The variation of this resistance with the ambient temperature helps to compensate the thermocouple for the cold junction temperature change. Some units also use a bimetallic compensator to cut off part of the incoming radiation as the temperature of the housing increases.

The distance between the source of radiation and the pyrometer is important, since the radiation must be focused onto the receiving disk. In some units, the position of the lens or mirror is adjusted so the radiation is always focused onto the same point; others use a fixed focus which requires that the radiation completely fill the pyrometer opening.

There is another type of instrument, the partial-radiation pyrometer, which uses photoelectric principles. It has some advantages over the total-radiation pyrometer, which can be sensitive to smoke, water vapor, and CO_2. The partial-radiation pyrometer can be adjusted to filter out part or all of the radiating effects of these mixtures. It can be used to measure the upper temperature of steel furnaces, where a total-radiation pyrometer would be in error due to the radiating properties of the water vapor and CO_2 present.

The use of a photocell in the partial-radiation pyrometer has an advantage in that variations in emissivity will not introduce as large an error as they do in other radiation pyrometers. However, variations in ambient temperature can introduce a larger error and partial-radiation pyrometers must be jacket insulated.

Single-band-pass pyrometers operate over a select, narrow band of the energy spectrum centered at a desired point. For the high-temperature measurement of metals, the band might be centered at the 0.65-μm point, which is the red end of the visible spectrum where metal emissivity is highest. In this case the instrument might be referred to as a brightness pyrometer.

The ratio or two-color pyrometer measures the energy from two rather narrow bands and divides one into the other. When the two bands are selected so that there is little change in emissivity between them (the bands are close together), the emissivity factor almost cancels out. The low-emissivity bodies that create most of the error for the broad- and single-band pyrometers do not have a large effect on the ratio type of instrument.

Emissivity over a narrow wavelength and band will not vary as much as it would over the total spectrum, but the limited-band pyrometer may have a lack of sensitivity because of the reduced energy available. The selection of the two wavelengths is made for the particular application.

The broad-band pyrometer is used generally for automatic control. It can cover all temperature ranges and is the least expensive of the three types. Narrow-band and two-color pyrometers are used to minimize the emissivity effects and for those applications where it is desirable to select a particular band pass.

The selection of a radiation pyrometer requires the consideration of the following:

1. Target temperature and limits.
2. Target material and emittance.
3. Target size and distance.
4. Angle of observation.
5. Target movement, speed of response of the pyrometer.
6. Atmospheric conditions between target and detector.
7. Sighting method, directly on target or through window required for vacuum or pressure.
8. Ambient temperature.
9. Scale temperature units.

The main advantages include

1. No physical contact with material being measured.
2. Fast speed of response, can be used on moving targets.
3. Can be used to measure small targets ($\frac{1}{16}$ in. or 1.6 mm in diameter) or to measure the average temperature over a wide area.
4. Can measure much higher temperatures than thermocouples and most other sensors.

The disadvantages include

1. More fragile and costlier than thermocouples, RTDs, or thermistors.
2. Emissivity of the target may cause a low temperature reading if corrections are not made.
3. Nonlinear scale shape, approximating the fourth power of the temperature.
4. A relatively wide temperature span is normally required.

REDUCING ERRORS IN TEMPERATURE MEASUREMENT

The use to which a temperature sensor is put can affect its accuracy in many ways. The successful application of the sensor thus depends on a knowledge of the errors involved. The error sources discussed below are descriptive of the problem areas most commonly found. Most of these errors are common to all of the temperature sensors that we have considered.

Calibration errors are a result of the uncertainty of calibrating the instrument at a known temperature. These are usually considered as a random error. Accuracy is a function of sensor configuration and calibration techniques and may range as low as 0.01°C at the ice point to 5°C at 1100°C.

Stability is normally a function of the environment. It refers to long-term changes in the calibration of the sensor. The output emf of a thermocouple will decrease with time at higher temperatures, while the resistance of a wire-wound resistance sensor increases.

When enough data can be obtained on a particular sensor from periodic calibrations, the changes can be classified as a systematic error and we can define them as

$$T = \frac{dT}{dt} t,$$

where

T = temperature error,
dT/dt = temperature drift rate,
t = time between calibrations.

Repeatability is the ability of the sensor to repeat a calibration at a known temperature. This tends to be a function of the sensor design since hysteresis becomes a factor in the calibrations. Repeatability can be better than 0.25°C over 500°C for some platinum resistance sensors. Since the hysteresis effects cannot usually be predicted, we consider repeatability as a random error.

Self-heating errors can occur from I^2R or Joule heating in a resistance. This causes the indicated temperature to be greater than the actual temperature. The self-heating error T_S can be expressed as

$$T_s = Q(R_i + R_b),$$

where

Q = power dissipated by the element,
R_i = internal thermal resistance of the element,
R_b = thermal resistance of the boundary layer.

The classical approach requires the solution of the steady-state heat transfer problem using an internal heat source, but it is usually easier to measure the self-heating effect directly.

One technique that can be used to measure the self-heating effect, provided that the temperature does not change rapidly, is to measure the resistance at several currents to obtain a number of measurable resistance changes. The I^2R power versus the change in resistance R or change in temperature T should be nearly linear. Self-heating effects can range from 0.02 mV/°C in some thermistors in still air to 500 mV/°C in platinum sensors in flowing fluids. These effects can be reduced by using a lower excitation current.

Stem conduction errors are due to the heat transfer to or from the sensing element along the sheath and leads. This results in the temperature of the sensor being different from that of the fluid. Stem conduction effects can be measured directly by incrementally withdrawing the sensor and measuring the temperature at different positions as discussed earlier.

A thermal lag error can occur when the temperature sensor does not respond fast enough to the environmental temperature changes. This error depends on the heat transfer characteristics of the environment as well as the sensor. If we know the rate of the environmental temperature changes and the sensor response time, then the error due to thermal lag TL can be found as shown below:

$$\text{TL} = \frac{dT}{dt} T_R,$$

where

dT/dt = rate of temperature change,
T_R = sensor response time.

Frictional heating occurs when a high-viscosity fluid flows around the surface of a sensor at high velocities. The temperature indicated by a sensor in such a flowing stream may be higher than the actual stream temperature. The frictional heating effects are a function of the square of the velocity, and a properly shaped shield can be used to minimize this effect.

Thermal emfs which develop in resistance elements and lead wires can also degrade sensor accuracy as discussed earlier. In thermocouples, the errors can easily be 0.1°C. The effects in resistance sensors using dc excitation are much less. A 100-ohm platinum sensor with a sensitivity of 0.5 mV/°C and 1 μV of thermal emf can be in error by about 0.002°C. Thermal emf errors can be reduced by maintaining the dissimilar metal joints at an equal temperature and the lead wires at the same temperature gradient. Alternating-current excitation can further eliminate these effects.

Thermal radiation errors can occur if the sensor receives radiation from a source which is at a higher temperature than the environment around the sensor. This error tends to be a function of the shape of the sensor with respect to the environment as well as the thermal radiation characteristics of the sensor. Thermal radiation errors can be minimized by the proper placement and shielding of the sensor.

The electrical insulation of a sensor can be a source of errors when in the presence of ionizing radiation, moisture, or high temperatures. In some temperature sensors, the degraded insulation will act as a resistor in parallel with the sensing element. This effect tends to lower the indicated temperature. The insulation resistance of an isolated sensor can be determined by measuring the resistance between the element and ground. Errors of this nature can be minimized by using better insulating materials.

Nuclear radiation effects include gramma rays, fast neutrons, and thermal neutrons. Radiation heating is the heat generated by the absorption of nuclear radiation. A sensor which dissipates 100 mW/°C can be in error by 1°C with 100 mW of gamma heating.

Fast neutrons cause dislocations in metals which will alter the sensor characteristics. These effects are worst at cryogenic temperature since at room temperatures the defects anneal out as quickly. The error for a platinum thermometer at $-253\,°C$ is about 1°C for a dose of $2 \times 10^{17} nvt$. Thermocouples can show errors about five times greater in similar environments.

Thermal or slow neutrons change the composition of materials by transmutation. Platinum is transmuted to gold and mercury, which changes resistance/temperature characteristics. This change is dependent and irreversible. A platinum thermometer will shift by about 1°C after a dose of $10^{22} nv_0 t$. A Pt/Pt-10Rh thermocouple can shift about 10°C.

Bridge errors also affect sensor accuracy. Power supply errors can occur as a result of the long-term stability drift of the power supply. Resistor aging is not usually a problem if stable wire-wound resistors are used. A shift of less than $\pm 0.01\%$ over a five-year time period is possible. Potentiometer shifts and noise can be caused by shocks and vibration. The film type of potentiometer may drift with age. Restricting the required adjustments will keep potentiometer errors to a minimum.

Temperature coefficients will not normally affect bridge resistors unless subjected to outdoor environments. Resolution errors in the form of steps can occur when wire-wound potentiometers are used.

Systematic errors have a direction and magnitude which may be predicted while random errors have a magnitude and direction which cannot be predicted. There is always some amount of randomness in systematic errors. This may be about three times the standard deviation of a sample of measurements. If the individual errors can be estimated, the random error can always be calculated as the square root of the sum of the squares of the individual errors.

A passive resistance bridge operating from a constant voltage supply is inherently nonlinear. This is due to the decrease in current that takes place when the sensor resistance increases. The use of a constant current supply is not completely successful since neither nickel nor platinum resistors are linear. The nonlinearity may be removed by scaling or conditioning or in some automation operations it may not be significant. A nickel resistor can be linearized by simple bridge techniques but for platinum active circuitry is required.

The linearization can be done using a microcomputer. The millivolt bridge signal is digitized and fed into the microcomputer, which computes the temperature.

48 TRANSDUCERS FOR AUTOMATION

In an automation system with a large number of temperatures at different ranges, the computation needs can grow quickly and the machine could become overburdened. The different constants require memory space and each measurement requires the multiplication of at least two constants plus squaring the voltage and adding another constant.

Another technique is to use feedback from the sensor leg of the bridge to modulate the bridge power supply. By increasing the supply voltage as the temperature rises, the tendency for the output to decrease is overcome and a linear output is achieved. Temperature transmitters can also be used which produce linear outputs for various types of resistance sensors.

In some cases, the sum of the individual accuracy factors can result in a total error which exceeds the desired system accuracy. It may be necessary to reduce the total error to the required accuracy limits by one or more of the following methods:

1. Using in-system calibration techniques with corrections performed by data reduction.
2. Specifying the accuracy over a total error band rather than on an individual parameter basis.
3. Monitoring the environmental changes and correcting the data accordingly.
4. Artificially controlling the transducer environment to minimize these errors.

Many of the individual errors may be random in nature. The total error will usually be less than the algebraic addition of these individual errors. The use of an error band tends to simulate actual conditions because only the total accuracy deviation is considered, rather than the individual factors contributing to it. Some individual errors are predictable and can be calibrated out. When the system is calibrated, the calibration data can be used to correct the recorded data.

Environmental errors can be corrected by data reduction methods if the environmental effects are recorded simultaneously with the data. Then the data can be corrected using the environmental characteristics of the transducer.

These techniques may provide a significant increase in system accuracy. They are useful when individual errors are specified and when the error band specification is not used.

EXERCISES

1. Discuss the construction and operation of a resistance temperature detector.
2. In a resistance thermometer the resistance at 20°C is 100 ohms. It is 101.5 ohms at 25°C. Compute the resistance values at -100°C and at 150°C assuming a linear relationship.

TEMPERATURE MEASUREMENT 49

3. Compute the resistance of the temperature-sensitive element in Exercise 2 at 10°C assuming a linear relation within 10% between resistance and temperature.
4. In the application of a platinum coil resistance thermometer why is it best not to exceed a measuring temperature of 500°C?
5. Explain, with the aid of a diagram, how a thermoelectric emf of the order of 0.001 V can be balanced in a resistance temperature system.
6. Discuss the advantages and disadvantages of using a thermocouple compared to a resistance temperature detector. Discuss such features as distance, speed of response, multipoint installation, accuracy, operating range, automatic calibration possibilities, and relative expense and maintenance.
7. Discuss the problems in the use of multiple thermistors for temperature measurement.
8. Describe the operation and considerations for a transistor temperature sensor.
9. Discuss the use and operation of a typical integrated circuit thermometer in an industrial automation application.
10. A platinum/platinum rhodium thermocouple has an emf temperature relationship given by

$$e = -3.28 \times 10^{-1} + 8.28 \times 10^{-2}T + 1.5 \times 10^{-5}T^2,$$

where e is in millivolts and T is the temperature difference, in degrees Celsius, between the hot and cold junctions. (a) Calculate, to three significant figures, the emf if the cold junction is at 20°C and the hot junction is at 1200°C. (b) Calculate the sensitivity, in microvolts, required to indicate a change of 1°C at 1200°C with the cold junction temperature maintained at 20°C.
11. Define the following thermocouple-related effects: (a) Seebeck effect, (b) Peltier effect, (c) Thomson effect.
12. Discuss the importance of the use of various metals in a thermocouple circuit.
13. Testing of a copper-Constantan thermocouple resulted in the following equation:

$$e = a(T_1 - T_2) + b(T_1^2 - T_2^2).$$

If $a = 3.25 \times 10^{-2}$ mV/°C and $b = 4.75 \times 10^{-4}$ mV/°C, $T = 100$°C, and the cold junction T_2 is in ice, calculate the emf in millivolts.
14. The emf of an iron Constantan thermocouple is 44.50 mV at 600°C with cold junction temperature of 0°C. If the cold junction temperature changes to 6°C, what is the corresponding emf? Assume a straight-line relationship between temperature and emf with the emf varying with temperature as 8.0×10^{-2} mV/°C.
15. List the possible errors that may be caused by the variation of resistance both internally and externally in thermocouple and resistance temperature measuring circuits.
16. Describe the use of a compensating loop in a temperature measurement circuit.
17. Discuss the use of blackbody radiation in high-temperature measurements.

3
PRESSURE TRANSDUCERS

PRESSURE MEASUREMENT

Pressure transducers are used in many automation applications. In some applications pressure transducers are used to measure liquid levels.

Pressure increases with liquid depth but it can be affected by density variations or changes in composition due to temperature differences. Figure 3-1 shows two types of submersible pressure transducers.

Many electrical output pressure tranducers detect pressure using a mechanical sensing element. These elements may consist of a thin-walled elastic member such as a plate or tube which provides a surface area for the pressure to act upon (Fig. 3-2). When the pressure is not balanced by an equal pressure acting on the opposite side of this surface, the element deflects as a result of the pressure. This deflection is then used to produce an electrical output. When another separate pressure is allowed on the other side on the surface the transducer will measure differential pressure. If the other side of the surface is evacuated and sealed, absolute pressure is obtained. The transducer will measure gauge pressure if ambient pressure is allowed on the reference side (Fig. 3-3).

INDUCTIVE PRESSURE TRANSDUCERS

Inductive pressure transducers use the pressure to move a mechanical member which is then used to change the inductance of a coil. The inductance is a function of the relative motion of a core and the inductive coil as shown in Figure 3-4.

Inductive single-coil transducers are sometimes used in oscillator circuits, where they are used to control the oscillation frequency. Single-coil transducers suffer from problems in compensating for temperature effects. This requires a matching of the core and winding materials for temperature versus permeability changes.

A more common type of inductive transducer uses the ratio of the reluctance of two coils. These reluctance transducers are less sensitive to temperature effects than the single-coil types. A motion of about 0.003 in. produces an ac output voltage of about 100 mV.

In a diaphragm-type reluctance pressure tranducer (Fig. 3-5), a diaphragm of magnetic material is supported between two symmetrical inductance assemblies.

(a)

(b)

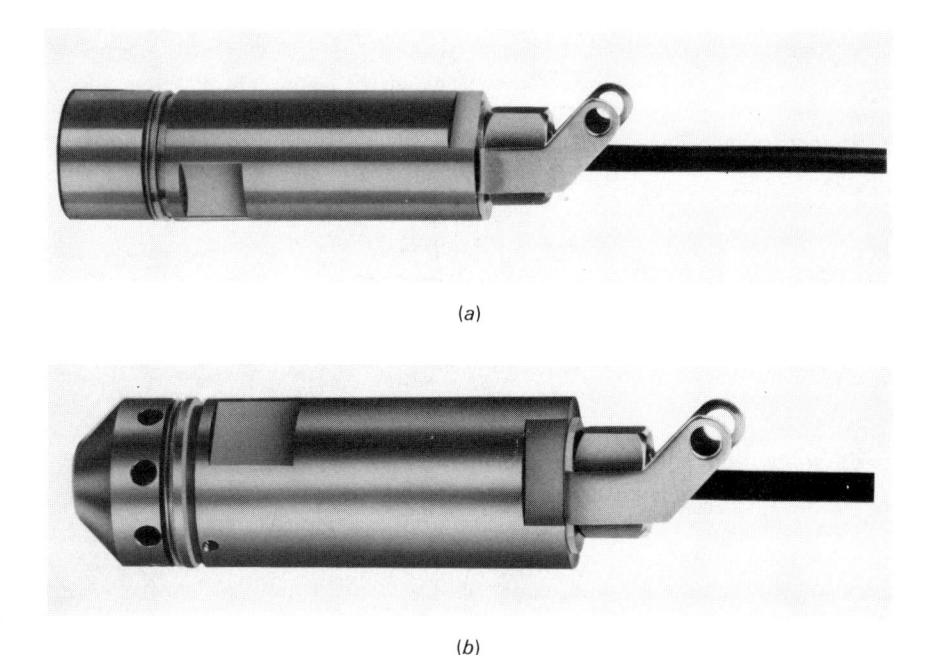

Figure 3-1. Submersible pressure transducers. Unit a may be submerged in liquid or attached to the vessel externally using a $\frac{1}{4}$-in. NPT connector. Unit b is designed for use in viscous liquids and slurries. (*Courtesy AMETEK, Inc.*)

This diaphragm is deflected when there is a difference in pressure between the two input ports. The gap in the magnetic flux path of one core will increase; in the other core it will decrease. In this way the reluctance changes with the pressure. The net effect is a change in inductance of two coils of the transducer. The inductance ratio L^1/L^2 may be measured in a bridge circuit to detect the voltage proportional to the pressure difference. The measurement of force due to the external pressure is thus accomplished by the change in the inductance ratio of the coils. The force being measured changes the magnetic coupling path of the transducer as a result of the displacement of the core.

The hysteresis errors in this type of transducer are limited to the mechanical components. The shape of the force summing members is governed by those configurations which are best suited to operate with the inductive principle. An E-shaped core is often used in order to maintain good balance and low phase shift.

The diaphragm can also be used as a part of the inductive loop. Here the overall performance of the transducer may be compromised since the mechanical characteristics of the diaphragm must be less than optimum to improve the magnetic characteristics.

52 TRANSDUCERS FOR AUTOMATION

Diaphragm

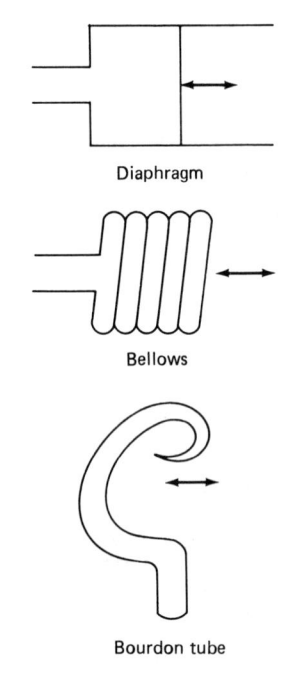

Bellows

Bourdon tube

Figure 3-2. Pressure sensing elements.

Differential pressure — Reference pressure

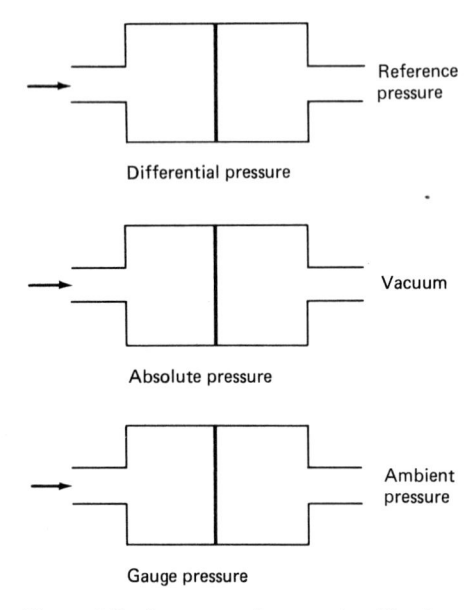

Absolute pressure — Vacuum

Gauge pressure — Ambient pressure

Figure 3-3. Pressure reference classifications.

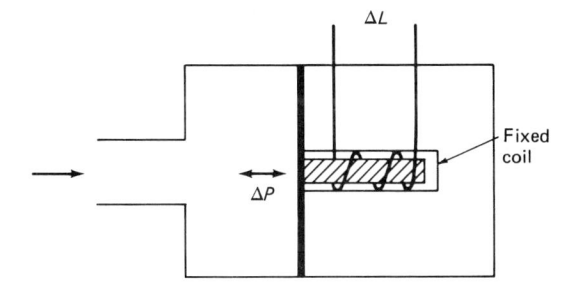

Figure 3-4. Single-coil inductive pressure transducer.

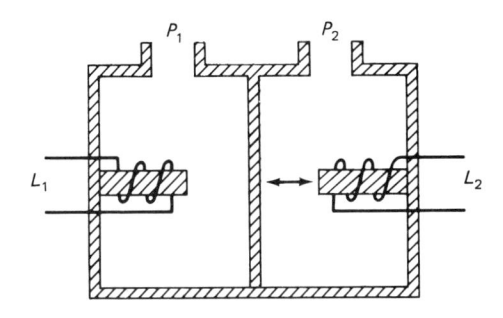

Figure 3-5. Differential reluctive pressure transducer.

This type of transducer provides the following advantages:

1. A high output is possible.
2. It can respond to both static and dynamic measurements.
3. Continuous resolution is provided.
4. A high signal/noise ratio is possible.
5. It can provide direct FM for telemetry connections.

Disadvantages include:

1. The frequency response is normally limited by the mechanical construction.
2. The excitation must be ac.
3. The transducer must be reactively and resistivity balanced at null.
4. Magnetic objects and nearby fields can cause transient errors.
5. The volumetric displacement tends to be large.
6. Mechanical friction can cause wear and errors over a period of time.

Some more recent transducers contain built-in dc to ac to dc conversion circuitry. Transducers with dc excitations of 28 and 5 V are available for ab-

54 TRANSDUCERS FOR AUTOMATION

solute, gauge, and differential pressure measurements. Range is typically one inch of water up to 12,000 psi.

The ac transducers normally use a carrier frequency in the range of 60 Hz to 30 kHz. When dc conversion circuitry is used, the internally generated carrier frequency can be much higher. This allows smaller coils and produces a more compact package.

The performance of these inductive transducers, with or without dc to dc conversion, can be on a par in many ways with the best versions of other pressure transducers. The static error is typically $\pm 0.5\%$ with the nonlinearity accounting for the major part of this. Errors due to hysteresis and nonrepeatability may be less than 0.2%. Proof pressure or overrange ratings of greater than six times normal range are available. Temperature effects are minimized by using similar sensing element and coil materials. These errors can be 1% or 2% to 100°F.

Transducers without dc conversion circuits can operate to 350°F. The solid-state components of the dc conversion circuits normally limit the operating temperature of this type of transducer to less than that of the ac types.

The frequency response range is 50 to 1000 Hz and it depends on mechanical design. Some particular constructions like the diaphragm types have a reasonable tolerance to shock and vibration and good pressure overload capabilities in liquid as well as gas systems.

The linear variable differential transformer (LVDT) type of construction as shown in Figure 3-6 uses a sliding core which is connected to the pressure sensing element. LVDT transducers are used widely for displacement velocity measurement. The LVDT may also be used to detect the movement of a bourdon tube (Fig. 3-7).

PIEZOELECTRIC PRESSURE TRANSDUCERS

Piezoelectric pressure transducers use a crystal to generate a charge or voltage when it is mechanically stressed. A diaphragm normally reacts with the pressure to produce the stress on the crystal. Piezoelectric pressure transducers can op-

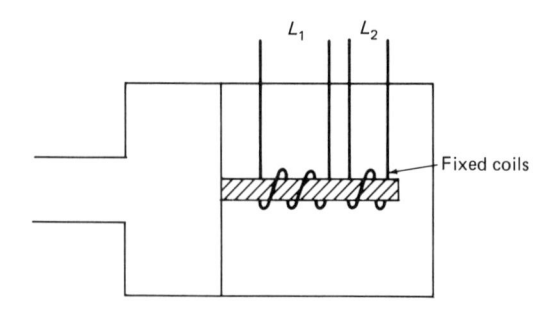

Figure 3-6. LVDT pressure transducer.

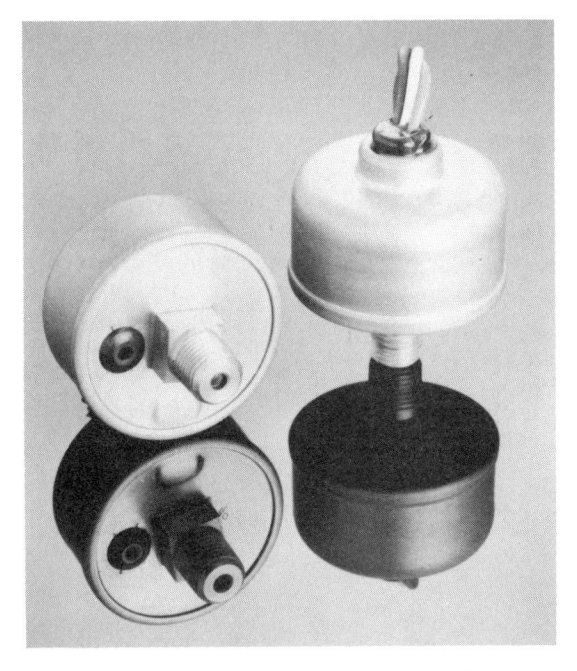

Figure 3-7. A sealed pressure sensor based on bourdon tube/LVDT technology. (*Courtesy AMETEK, Inc.*)

erate over a wide temperature range with relatively small errors due to temperature changes.

A number of crystal materials are used. Some quartz units use crystals selected from those found in the natural state. The grown crystals include ADP (ammonium dihydrogen phosphate) and are grown in an aqueous solution. Various ceramic materials are also used.

The piezoelectric elements are cut from the crystal along the existing crystallographic axes. Ceramic elements are pressed from powdered materials into the required shape and then fired at high temperatures. The piezoelectric characteristics result as these are polarized by an electric field during the cooling process.

Ceramic elements may reach the Curie point when heated. This is the temperature at which the crystalline structure changes and polarization is lost. The element then ceases functioning. Curie points vary from 300°F to over 1000°F. The element must then be polarized. Quartz elements can be used at temperatures ranging from −400 to 500°F.

The output of piezoelectric elements can also be affected by the pyroelectric effect, which causes changes in the output proportional to the rate of change of temperature experienced by the crystal.

56 TRANSDUCERS FOR AUTOMATION

Some quartz-crystal transducers have been used with amplifiers which permit static measurements, but most piezoelectric pressure transducers are used for the dynamic measurement of rapidly varying pressures. Sound pressure levels of 180 dB and more may be measured with an accuracy of 3% of full scale, including high-frequency components. Quartz units have been mounted as spark-plugs to measure pressures in internal combustion engines. Piezoelectric transducers are also used to sense the pressure surges due to shocks. They have been used on tubines, pumps, and hydraulic equipment to measure dynamic stresses.

Frequency response is in the range of 10 Hz to 50,000 Hz and the pressure ranges are 5 to 10,000 psi. Quartz and ADP crystal units generally have a higher natural frequency than the ceramic units; however, the ceramic crystals provide higher output levels.

Piezoelectric sensors respond well to high shock levels and can be a problem in some high-level applications. The high-level shock could produce an over voltage which would saturate the signal conditioning amplifiers for a period of 50 or more times the shock duration, causing a loss of data during this time.

The normal output of a piezoelectric crystal is low and the impedance high so a high-impedance amplifier is normally used for signal conditioning. An operational amplifier can be used with capacitive feedback to compensate for the capacitance due to cabling. Integrated circuit technology allows the amplifiers to be placed inside the transducer case on some commercial units.

CAPACITIVE PRESSURE TRANSDUCERS

Capacitive pressure transducers use a metal diaphragm for one capacitor plate with the other plate positioned alongside of the diaphragm as shown in Figure 3-8. The pressure causes a movement of the diaphragm which changes the capacitance between the two plates. An ac signal is placed across the plates to sense the change in capacitance.

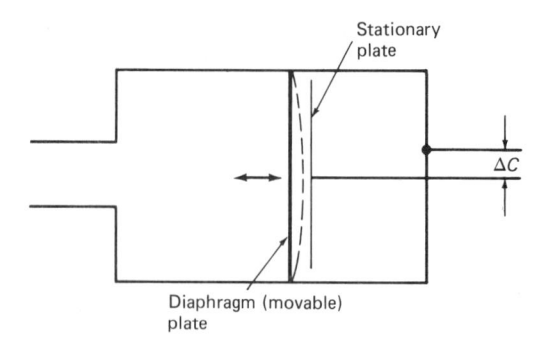

Figure 3-8. Capacitive pressure transducer.

The capacitive sensor can be used as part of an *RC* or *LC* network in an oscillator or it may be used as a reactive element in an ac bridge. When it is used in an oscillator circuit, the output may be ac, dc, digital, or in the form of a phase shift.

Capacitive transducers are small in size with a high frequency response. They can be operated at high temperatures and allow the measurement of both static and dynamic quantities. The range is 0.01 to 10,000 psi with a typical error of 0.25%. Units are also available with accuracies to 0.05%.

Capacitance transducers must be reactively as well as resistively matched. Long lead lengths or loose leads can cause a variation in capacitance due to capacitive coupling. It is desirable to use a preamplifier close to the transducer as well as matched cabling to minimize these coupling effects.

Some capacitive sensor designs use materials such as quartz for the capacitor plates. One unit uses two thin quartz disks with platinum electrodes on their inner surfaces. The disks are fused to form a small capsule and the electrodes separated by a 0.0002-in. gap. As a vacuum is drawn inside the gap, the capacitance changes and absolute pressures of up to 30 psi can be measured. Gauge pressures of up to 30 psi can be measured with one port vented to the atmosphere. The device provides an accuracy of $\pm 0.5\%$ as well as a 5-V TTL output.

The major disadvantages of capacitive sensors are

1. The high impedance output must be reactively and resistively balanced.
2. Movement of connecting cables or long lead length will cause distortion or erratic signals.
3. Most units are sensitive to temperature variations.
4. The receiving and conditioning circuitry can be complex compared to that for other types of sensors.

The major advantages are

1. The higher frequency response.
2. The actual sensor is inexpensive to produce and has a small volume.
3. A low shock response is possible due to the minimum mass required for the diaphragm.
4. Either static or dynamic measurements can be made.

POTENTIOMETRIC PRESSURE TRANSDUCERS

A potentiometric transducer is an electromechanical unit with a resistance element which is contacted by a movable slider. The motion of the slider due to a pressure change results in a resistance change as shown in Figure 3-9.

Deposited carbon, platinum film, and other resistive elements are used. The

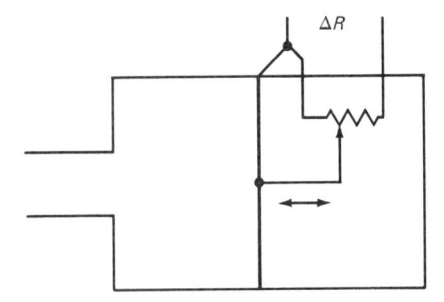

Figure 3-9. Potentiometric pressure transducer.

contact between the resistive element and the slider can be a major source of noise and errors due to inconsistent wiper pressure.

The potentiometer sensor is widely used despite its limitations. Its electrical efficiency is high and it provides an output for many operations without further amplification.

The major advantages are

1. High output.
2. May be excited with ac or dc.
3. Inexpensive.
4. No amplification or impedance matching may be required for transmission.
5. Wide range of output functions available.

The disadvantages are

1. The high mechanical friction can cause limited life.
2. The resolution is finite in many devices.
3. High noise levels can develop owing to contact wear.
4. Units are sensitive to vibration.
5. The accuracy is a function of the force required because of friction.
6. The frequency response is low because of the mechanical contact.
7. Large size.
8. Large displacements may be required from pressure sensing elements.

The first potentiometric pressure transducer was reported in 1914. They are still used today because of their low cost and connection simplicity. A wire-wound or deposited conductive film may be used.

The output due to a pressure change is a function of the design of the potentiometer resistance curve, which can be linear, sine, cosine, logarithmic, or exponential. The unit can be used with ac or dc and no amplification or impedance matching may be required. A high output can be obtained with a high input voltage.

Recent trends in potentiometric pressure transducers have gone in two directions: (1) Miniaturized devices with more relaxed tolerances. (2) The use of a control force to supplement the force of the diaphragm or capsule. These motor- or force-driven systems allow the use of low-resolution multiturn potentiometers for improved accuracy.

The more typical potentiometric transducers have an error of $\pm 1\%$. A pressure range of 5 to 400 psi is typical and high-pressure devices are available to 10,000 psi. Specialized devices are available with the following specifications: resolution 0.2%, linearity $\pm 0.4\%$, hysteresis 0.5%, and temperature error $\pm 0.8\%$. Some advanced instruments offer $\pm 0.25\%$ end-point accuracy.

STRAIN GAUGE PRESSURE TRANSDUCERS

Strain gauge pressure transducers use the force from a pressure change to cause a resistance change due to mechanical strain. Figure 3-10 shows several different package types. The pressure sensing element may be a diaphragm or even a straight tube since the deflection required is small.

If a tube is used, the strain gauges can be mounted right on the tube, which is sealed at one end. The pressure difference causes a slight expansion or contraction of the tube diameter Table 3.1 gives the characteristics of strain gauge transducers.

Some designs use a secondary sensing element or auxiliary member in the form of a beam or armature. Four or two arms of a Wheatstone bridge may be used for temperature compensation.

The force being measured displaces and changes the length of the member to which the strain gauge is attached. The strain gauge property known as the gauge factor produces a change in resistance proportional to the change in length. The strain gauges may be arranged in the form of a Wheatstone Bridge circuit with one to four of the bridge legs active.

Strain gauge transducers may be unbonded or bonded. An unbonded gauge has one end fixed while the other end is movable and attached to the force member. The bonded gauge is entirely attached by an adhesive to the member whose strain is to be measured (Fig. 3-11). Strain gauges may be made from metal and metal alloys, semiconductor materials, or thin-film materials.

The gauge factor property that converts mechanical displacement is defined as the unit change in resistance per unit change in length. Although all electrical conductors have a gauge factor, only a few have the necessary properties to be useful as strain gauges. For the more common metal strain gauge materials, the gauge factor ranges from 2.0 to 5.0. Higher-gauge-factor materials tend to be

60 TRANSDUCERS FOR AUTOMATION

(a)

(b)

Figure 3-10. These strain gauge pressure sensors use silicon gauges arranged in a bridge configuration. The strain gauges are diffused from a single semiconductor chip and laser trimmed for temperature compensation. a. Die-cast sealed unit with 1/8-27 NPT connection. b. Air pressure unit which can be mounted directly on a circuit board. c. Miniature unit with 9/16-18 UNF connection. (*Courtesy AMETEK, Inc.*)

(c)

Figure 3-10 (*Continued*)

Table 3.1. Strain gauge pressure transducer characteristics.

Design pressure—To 200,000 psig (1400 MPa)
Design temperature—Typically to 250°F (120°C)
                            special designs to 600°F (316°C)
Materials—Stainless steel and other corrosion-resistant metals
Error—±0.1 to ±1% of total span
Range—3 in. H_2O to 200,000 psig (0.08 kPa to 1400 MPa)

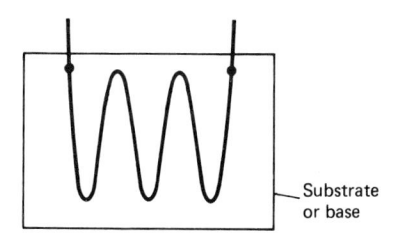

Figure 3-11. Wire strain gauge, bonded type.

more sensitive to temperature and less stable than low-gauge-factor materials. The strain sensing filaments must have stable elastic properties, high tensile strength, and corrosion resistance. The alloys used must have the proper combination of gauge factor and thermal coefficient of resistivity to give optimum performance.

62 TRANSDUCERS FOR AUTOMATION

If the strain gauge filament functions as the four elements of the bridge circuit, it is divided into four parts equal in resistance value. A resistance of 350 ohms is typical, but bridge resistances from 50 ohms to several thousand ohms can be achieved by varying the diameter of the wire, the length of the filaments, and the number of loops of the resistance wire wrappings. This versatility allows the unbonded type of strain gauge transducer to be used in a wide variety of systems and automated equipment.

In a typical unbonded unit, a stationary frame is used with an armature. Four filaments of strain-sensitive resistance wire are wound between rigid insulators which are mounted in the frame and armature. The filaments are of equal length and are arranged as shown in Figure 3-12. The strain-sensitive resistance wire must be wound to keep the filaments under some residual tension when the mounting posts are displaced to either end position.

The resistance of the four strain-sensitive elements is trimmed during assembly such that no signal appears in the output circuit when there is not an external force. If the bridge is balanced for zero output, the unbalanced electrical output of the bridge will tend to have a linear relationship to a force applied.

In some units the mechanical arrangement is such that the travel of the armature can be as small as ± 0.0003 in. This amount of movement is available in many of the miniature transducers. These units provide excellent frequency response because of the low displacement and small mass used in the armature.

In a bonded strain gauge the electrical insulation is provided by an adhesive or insulating material on the strain gauge. The force needed to produce the displacement is larger than that required with an unbonded strain gauge because

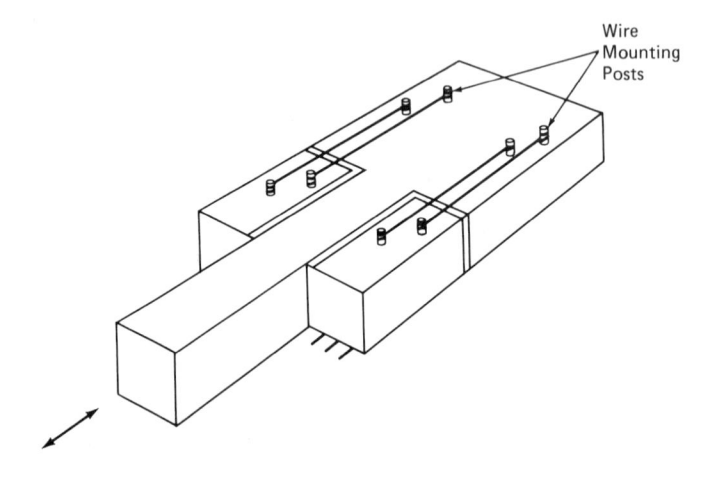

Figure 3-12. Strain gauge with armature, unbonded.

of the additional stiffness. The bonded strain gauge member can be used in a tension, compression, or bending mode.

Semiconductor materials such as silicon can also be used for strain gauges. By changing the amount and type of dopant, the strain gauge properties can be modified for a specific application. Silicon strain gauges have a higher gauge factor than metal strain gauges, but they also have a higher temperature coefficient. Gauge factors of 50 to 200 are typical and they may be either positive or negative.

The types of available strain gauge elements used in pressure transducers include

1. Unbonded metal wire gauges.
2. Bonded metal wire gauges.
3. Bonded metal foil gauges.
4. Thin-film deposited gauges.
5. Bonded semiconductor gauges.
6. Integrally diffused semiconductor gauges.

Unbonded wire elements (Fig. 3-12) are stretched and unsupported between a fixed and a moving end. These sensors tend to have high sensitivity, but they are sensitive to vibration. The unbonded strain gauge can be used for pressures of less than 5 psi. Nichrome and platinum wire are commonly used.

Bonded elements are attached permanently to the active strain element. This makes them much less sensitive to vibration. The cut or etched foil types allow a strong bond as well as more automated trimming. The foil, thin-film, and semiconductor strain elements are normally bonded or deposited such that the semiconductor, thin film, or foil and a pressure diaphragm appear as a single part.

Thin-film strain gauges use manufacturing techniques which are similar to those used for electronic microcircuits such as vacuum deposition. By controlling the materials and deposition processes, strain gauge properties can be modified to produce the desired characteristics.

In these transducers a metal substrate provides the desired mechanical base; next a ceramic film is vacuum deposited on the metal to provide electrical insulation. Four strain gauges are then vacuum deposited on the insulator. They are connected into a bridge configuration by vacuum-deposited leads. These multiple evaporations can be made during a single vacuum pumpdown period by using multiple sources and substrate masks.

The leads are attached to the film by the microcircuit techniques of microwelding or thermocompression bonding of the noble metal wire. The lead wire attachment is made directly to the film. The sensing element can take on many configurations since the strain gauges can be deposited on diaphragms, beams, columns, and other elements.

The strain gauge pattern can be designed to optimize the characteristics of a specific sensing element. A common pressure element is the flat-plate diaphragm with four strain gauges arranged in a bridge configuration.

When a force is applied normal to the plate, the plate is deformed, causing the strain-sensitive elements to elongate and increase in resistance. The change in resistance is proportional to the change in length and alters the balance of the bridge to produce an electrical output.

Diffused semiconductor strain gauges are built using the same integrated circuit technology. The pressure sensing element is produced by diffusing a four-arm strain gauge bridge on the surface of a single-crystal silicon diaphragm with a diameter and thickness selected for the pressure range and application.

The silicon diaphragm has the required pressure sensing mechanical properties since it is elastic and free from hysteresis effects. These sensors have high gauge factors and produce relatively high outputs at low strain levels. The transducer is encapsulated in a housing using IC manufacturing techniques such as electrostatic or thermal compression bonding or electron beam welding.

Such operational ratings as shock, vibration, and overload are similar to those of other high-quality microcircuit devices. The combined linearity and hysteresis effects are less than 0.06%. The low mass of the silicon diaphragm gives a fast response with minimum sensitivity to accelerations due to shock and vibration. In those applications where the media are not compatible with silicon, stainless steel, Hastelloy, or another material is used to isolating the diaphragms.

The full-scale output of a four-element strain gauge bridge for a pressure transducer using metal wire or foil gauges is 50 to 60 mV for bonded gauges 60 to 80 mV for the unbonded gauges, using 10-V excitation. Compensating and adjusting resistors may reduce the output to about 20 to 30 mV for bonded and 30 to 40 mV for unbonded gauges. These resistors are used for zero and balance adjustments, full-scale adjustment, thermal zero shift compensation, thermal sensitivity shift compensation, and shunt calibration.

Semiconductor strain gauge transducers provide an output of 200-400 mV for 5-7 mA excitation. Many transducers have a limitation on the excitation voltage to prevent destructive heating. Internal circuitry may be used to reduce the voltage to the level required for the proper bridge operation. A constant current source may be utilized to provide thermal compensation.

In order to provide TTL output levels, an amplifier must be used for all metal and some semiconductor strain gauge transducers. Many diffused circuit transducers use integral amplifiers, which may also contain some thermal compensation functions.

Strain gauge transducers provide fast response times, good resolution, minimal mechanical motion, and good accuracy. Predictable compensation methods for temperature effects, a low source impedance, and freedom from acceleration effects for the bonded type are among their other advantages.

Obtaining a zero output at zero pressure due to bridge unbalances can be difficult and there may be high vibration errors when unbonded types are used, especially for ranges lower than 15 psi. The low output levels can also result in noise problems, and usually isolation of the excitation ground from the output ground is required in addition to other signal conditioning requirements.

PRESSURE TRANSDUCER ERRORS

Ambient temperature variation is a major source of errors in many pressure transducers. These errors can be of two types: (1) The zero setting may shift with temperature. This can be due to unequal mechanical expansion of the instrument members. (2) The calibration factor or sensitivity may change with the ambient temperature. This may be caused by a change in elasticity or spring constant of the gauge members. Many metals have a temperature coefficient for Young's modulus of elasticity of about $-0.0007°C$. In order to minimize the zero error with temperature, the differential expansion of the mechanical components must nearly balance; otherwise the armature of the transducer can be pulled off center and the range of span becomes incorrect even if the zero shift is corrected for.

This differential expansion must be reduced until the total change over the ambient temperature range is a small fraction of full scale; then the remaining error can be compensated electrically as shown in Figure 3-13. Other variations are possible using circuits with compensating properties.

The compensating resistor should be at the same temperature as the transducer and preferably inside the case. The compensating resistors should not dissipate any appreciable heat or an allowance must be made for the increased resistance from this cause. The temperature coefficients of the compensating resistors' materials can be found from conducting tests of the actual resistors to be used. A number of factors can affect the resistance of a particular type of resistor and the only positive method of obtaining the coefficient is to conduct a test to duplicate the conditions of service.

The selection of the appropriate transducer is the first and most important step in obtaining accurate measurements. The transducer has the critical function of transforming the measurand from a physical quantity to a proportional signal. The accuracy of the final data can never be any better than the transducer's capability.

66 TRANSDUCERS FOR AUTOMATION

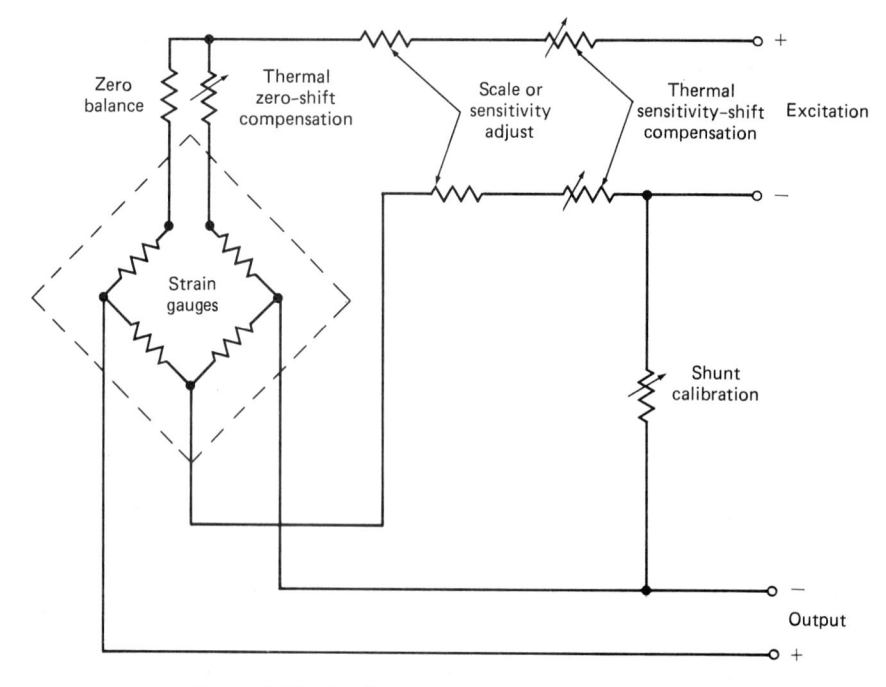

Figure 3-13. Strain gauge bridge with compensation.

A useful technique for improving system accuracy is to control artificially the environment of the transducer. If the local environment can be kept unchanged, these errors will be reduced to zero. This type of error control may require physically moving the transducer or providing the required isolation from the environment by a heater, thermal insulation, or similar means.

The force-balance technique is used in some pressure transducers to reduce errors. In these pressure sensing systems, the output of the sensing device is fed to an amplifier, which in turn feeds back a restoring force equal to the applied variable.

The sensing device is thus returned to the position it occupied prior to the application of the force being measured. The magnitude of the signal fed back from the amplifier determines the output of the system. The repositioning of the sensing head may be accomplished by motors or electromagnetic force.

The force being measured may be sensed by almost any of the techniques discussed in this chapter. Many of these units use a capacitive or differential transformer sensing element.

In some types of force-balance systems optical techniques are used. The pressure sensor in one type of manometer system uses a hollow spiral quartz tube with two wire-wound coils suspended from it. The coils are in a permanent magnetic field and a curved mirror attached to the tube transmits the tube motion to an optical detector.

Other high-accuracy systems for pressure measurement rely on the accuracy and readability of digital readouts. These may use displacement producing sensors such as bellows or diaphragms as the primary sensing device with analog-to-digital converters.

One technique uses a bourdon coil pressure sensing element and an optical encoder. These may be mounted on a common shaft which rotates in proportion to the applied pressure. A direct binary BCD or cyclic binary output is provided in serial or parallel form to displays (Fig. 3.14). The digital format is compatible with microcomputer interfaces for error reduction processing, monitoring, and recording.

Digital pressure instruments form an important part of pressure measuring equipment and their use is growing. High-performance digital instruments are available which can be used to calibrate other pressure transducers or gauges

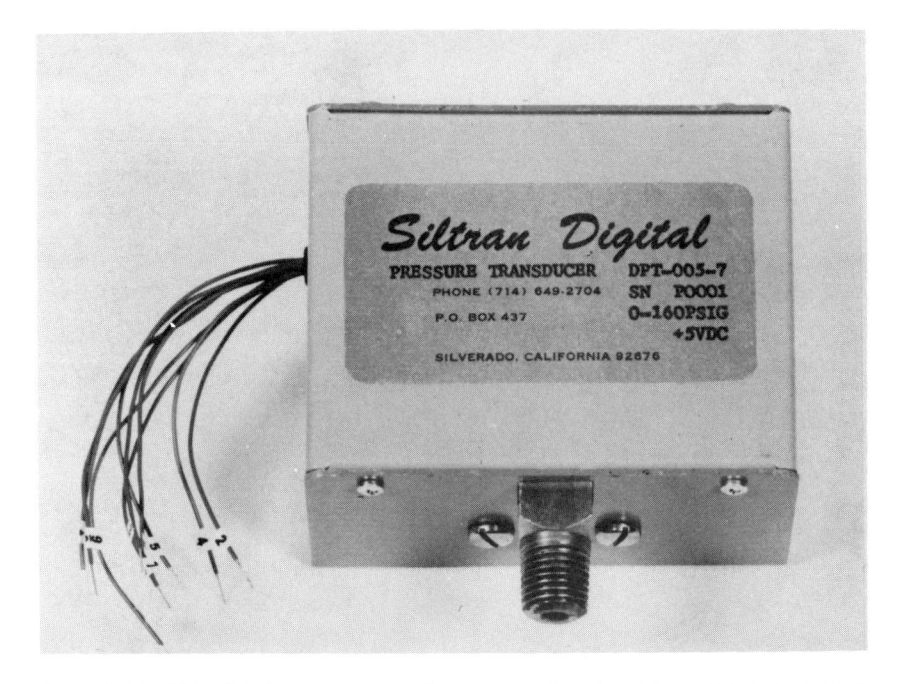

Figure 3-14. This digital pressure transducer uses a bourdon tube which is tracked by an optical encoder. The 8-bit output is parallel cyclic binary. (*Courtesy Siltran Digital*)

68 TRANSDUCERS FOR AUTOMATION

(a)

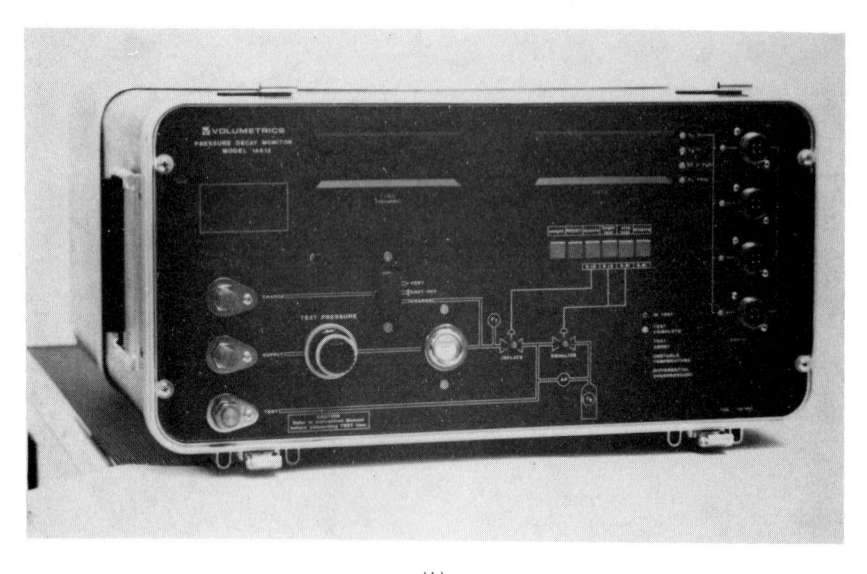

(b)

Figure 3-15. Digital pressure instruments for calibration. *a*. Microprocessor-based unit with ±0.05% FS accuracy traceable to NBS. *b*. Microprocessor-based leak detector, sensitive to 0.001 in. of water. (*Courtesy Volumetrics*)

(Fig. 3-15). Digital instruments have the flexibility to provide the pressure display in almost any pressure units.

EXERCISES

1. Describe the characteristic problems of using a potentiometric pressure transducer.
2. Discuss the advantages of using a capacitive pressure transducer over a potentiometric transducer in an automatic control system.
3. What pressure is obtained if the reference side of a sensor is evacuated?
4. What are the advantages of a semiconductor strain gauge over a metallic foil strain gauge in a high-volume manufacturing application?
5. What are the characteristics of piezoelectric pressure sensors that make them more suitable over other types in a manufacturing application? Discuss some typical automation applications.
6. The pressure of liquid oxygen flowing through a small diameter tube is to be determined. Required accuracy is ±1%, range is 20 to 80 psi, time constant is 0.8 s, proof pressure of 160 psig is required, flow rate is 8 to 15 gpm. Design a monitor and alarm system for a series of four similar stations with outputs to a microcomputer located at a maximum distance of 120 ft. Design the system for maximum flexibility and reliability.
7. A 300-ohm strain gauge with a gauge factor of 4.8 is to be measured using a 50-ohm cable in a bridge circuit having matched gauge resistors. What current will flow for a 1800-μin./in. strain when 10 V is applied to the circuit?
8. The circuit for a transducer is a complete and balanced Wheatstone bridge. The power supply is to be either alternating current or a carrier system. Discuss the choices and considerations involved in interfacing this to a microcomputer.
9. If a pressure of 9.0 in. of water is acting on an effective area of 2.0 in. of a diaphragm, calculate the force, in pounds, acting on the transducer push rod member.
10. Compare the advantages and disadvantages of bonded and unbonded strain gauges.
11. Describe the operation of a diffused diaphragm strain gauge pressure transducer.
12. If the approximate effective area of a diaphragm is given by $A = (R_1 + R_2)/2$, calculate the effective force acting on a central push rod if a differential pressure of 8.0 psi is applied to the unit and $R_1 = 1$ in. and $R_2 = 1\frac{1}{4}$ in.
13. What are some of the factors that determine the deflection range of strain gauge pressure transducers?
14. Compare the physical differences and applications of foil thin-film strain gauge elements.
15. List the important physical characteristics of crystals for pressure transduction.
16. Describe the operation of the force-balance technique.
17. A sensitivity adjustment controls the ratio of output signal to excitation voltage per unit measured. Describe some ways this can be accomplished.
18. An error band is defined as the band of allowable deviations of output values from a specified reference line or curve. These deviations are attributable to the transducer. How can the error band be derived?

70 TRANSDUCERS FOR AUTOMATION

19. Select a pressure transducer with the following advantages: (a) high output, (b) high accuracy, (c) static or dynamic measurements can be made, (d) high stability, (e) high resolution. Justify your answer.
20. Diagram some techniques for temperature compensation in strain gauge pressure transducers.

4
FLOW TRANSDUCERS

INTRODUCTION TO DIFFERENTIAL PRESSURE FLOW TRANSDUCERS

The automation of almost every type of chemical process requires the measurement of flow. The medium may be a fuel, gas, water, air, or processed liquids or gases which form the product itself. In order to specify a flow sensor for such a measurement one must have a knowledge of the variety of devices available.

Some flow sensors respond directly to the flow rate of a fluid fall. These sensors can be classified into three general groups:

1. A restriction in a pipe or duct is used to produce a differential pressure which is proportional to the flow rate. The differential pressure is then measured using a pressure transducer system which is calibrated in flow units.
2. A mechanical member is used to sense the flow of a moving fluid by rotation or deflection in a tapered tube.
3. The fluid's physical characteristics are used to sense the flow.

The sensing elements in the first group are sometimes called head meters since the differential pressure between two sensing ports may be equated to the head or the height of the liquid column. These elements provide a constant area for the flow passage. Venturi tubes, flow nozzles, orifice plates, and Pitot tubes are all examples of this group. The pressure transducers used to measure flow with these sensing elements are described in Chapter 3.

The head meter type of flow measurement is simple, reliable, inexpensive, and can offer good accuracy. Since it is a direct measurement of flow rate, it is suited to automatic flow control systems.

ORIFICE PLATES

Orifice plates provide a simple restriction for the differential pressure measurements of liquids or gases. These restricting plates are mounted in the flow line, perpendicular to the flow. The flat orifice plate has an opening smaller than the

72 TRANSDUCERS FOR AUTOMATION

inside diameter of the piping as shown in Figure 4-1. Taps are made at two points in the line in order to measure the pressure upstream from the orifice and downstream at the point of lowest pressure. The pressure difference between the two taps is used to calculate the rate of flow.

Orifice plates are not practical for slurries and dirty fluids because of accumulations and wear at the orifice edges. The proper sizing of an orifice can be accomplished with tables which can include compressibility factors, Reynolds curves, and thermal expansion factors. However, to have an accurate sensing unit for most applications, it is more common to depend on the orifice manufacturer to furnish a proper orifice from a specification of the application.

The flow rangeability of a particular orifice plate tends to be low, on the order of 3 or 4 to 1. This means the flow may not change by more than a factor of 4. Range changes are effected by changing the plate size. There are also limitations as to the minimum line size and minimum flow rates for accuracy when orifice plates are used. The quadrant-edge orifice plate is used for low flow rates or for flows involving viscous fluids. The concentric-thin-plate square-edge orifice has the greatest application. Eccentric and segmental plates are used to a lesser degree. An accuracy of 1% is possible with careful installation.

The advantages of orifice plates include

1. Low cost of the restriction itself.
2. Capacity changes made by switching the plate size.
3. A large amount of available coefficient data is available.
4. They are applicable to a wide range of temperatures and pressures.

The disadvantages include

1. A straight approach of piping is required.
2. There is a low rangeability for a particular plate.
3. Upstream-edge wear causes inaccuracies to develop.
4. They are unsuitable for slurry applications or for low-pressure systems.

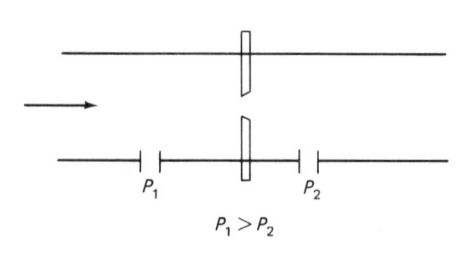

Figure 4-1. Orifice plate flow sensor.

VENTURI TUBES AND FLOW NOZZLES

Another type of flow sensing element which operates in a similar manner to the orifice plate is the Venturi tube. The Venturi tube as shown in Figure 4-2 combines a short, constricted portion called a throat between two tapered sections. It is usually installed between flanges in the line. In actual operation it tends to accelerate the fluid and lower its pressure.

Venturi tubes are normally used for liquids. They are used for gas flow measurement when the pressure recovery characteristics allow.

An orifice is a restriction at a point in the line, while the Venturi tube spreads the restriction over a longer distance. Flow through the center section is at a higher velocity than at the end sections. Taps at the entrance and throat measure the pressure difference. An accuracy of 1% is possible.

The advantages of Venturi tubes include

1. High capacity.
2. The pressure recovery is better than for orifice plates.
3. Fluids containing suspended solids can be used.
4. Resist wear due to abrasion.

The disadvantages are

1. The larger sizes require considerable room.
2. More costly than orifice plates.
3. They produce lower differentials than orifices for the same flow and throat size.
4. Low rangeability of 3 to 4 to 1—it is difficult to modify them for major changes in flow range.

Similar to the Venturi are the Dall and Foster tubes.

The flow nozzle is a curved nozzle as shown in Figure 4-3 which is inserted between two sections of piping. The curvature of the inside contour must ap-

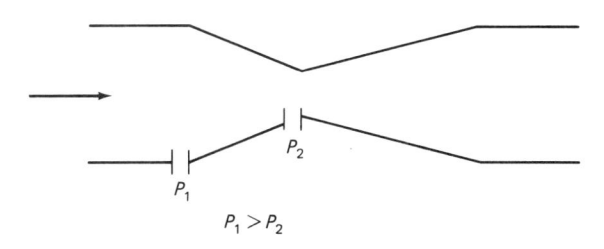

Figure 4-2. Venturi tube flow sensor.

74 TRANSDUCERS FOR AUTOMATION

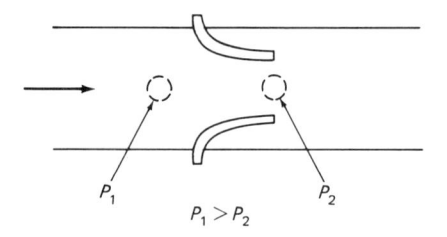

Figure 4-3. Flow nozzle sensor.

proach a gradual tangency to the throat without a sudden change of contour. The outlet end of the nozzle is usually beveled or recessed.

The advantages of flow nozzles are

1. They are suitable for fluids containing small amounts of solids.
2. They are more rugged than orifice plates and can be used for high-velocity measurement.
3. Their pressure recovery is better than that of orifice plates.
4. They have 65% greater capacity than orifice plates of the same diameter.
5. They are not as susceptible to wear as orifice plates.
6. They are more easily installed than Venturi tubes.
7. They tend to produce higher differential pressures than Venturi tubes.

The Pitot tube is not commonly used in industrial systems. It is sometimes used for spot-checking flows. A typical Pitot tube is shown in Figure 4-4. It has two pressure openings, one of which faces the flowing fluid. This opening intercepts a small portion of the flow and mainly reacts to the impact pressure.

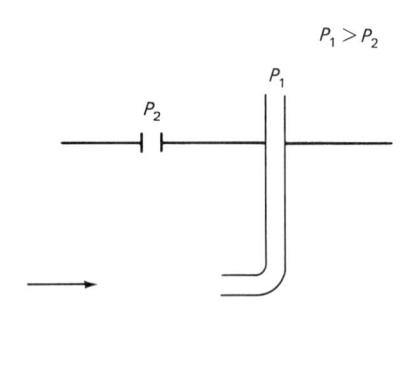

Figure 4-4. Pitot tube flow sensor.

The other opening is perpendicular to the axis of flow, it measures the static pressure.

There are other differential pressure sections in use such as elbows, which can be useful for problem viscous fluids, and loops, which tend to react well to pulsating flows and slurries.

TURBINE FLOWMETERS

The turbine flowmeter uses the movement of the moving fluid to turn a turbine wheel as shown in Figure 4-5. The speed of the rotor is proportional to the flow rate.

The turbine flowmeter output is a precise number of pulses which depends on the volume of the fluid displaced between the rotor blades in a unit of time.

As each blade passes the coil's pole piece, a voltage is induced in the coil. The transducer output is an ac voltage with a frequency proportional to the flow rate. The number of cycles per revolution is a function of the number of blades on the rotor and the revolutions per unit time are a function of the flow rate.

The rotor blades can be shaped so as to produce a sinusoidal output voltage with little harmonic distortion. The transducer output frequency band for a range of flow rates may be selected to coincide with a particular subcarrier frequency

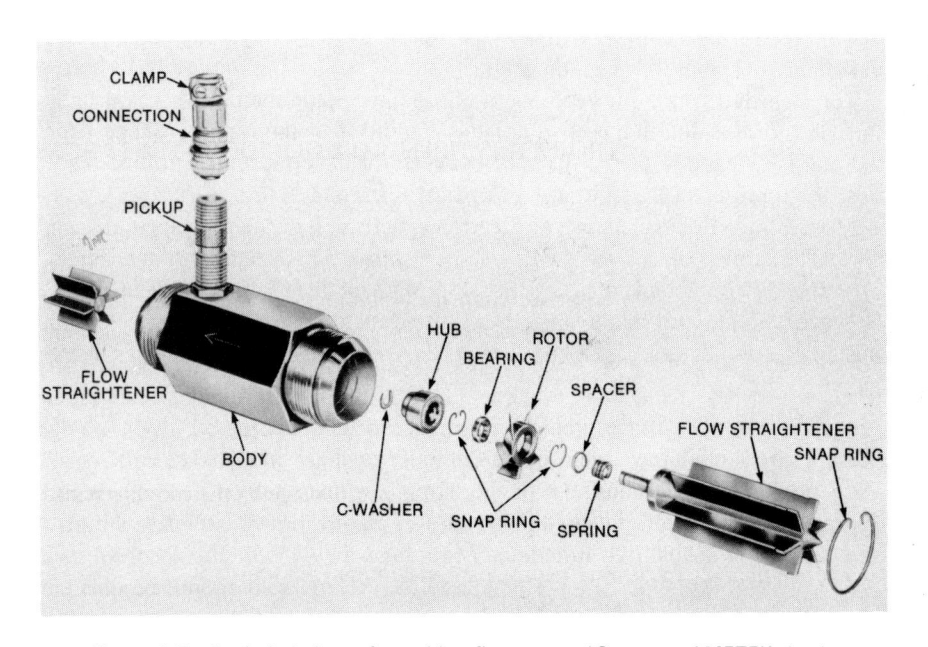

Figure 4-5. Exploded view of a turbine flowmeter. (*Courtesy AMETEK, Inc.*)

band for telemetry purposes. This is set by the number of rotor blades and their pitch.

Turbine meters have become widely used in many processing industries because of their wide range and performance characteristics. Turbine flowmeters can also be used for gas flow measurements.

Turbine meters should be selected with a capacity from 30 to 50% above the expected maximum flow rate. Operating these units below the maximum capacity provides a greater reliability.

Turbine flowmeters have a high performance. Each meter is individually calibrated for one or a number of fluid conditions before shipment. They are among the most accurate flow monitoring devices in common use.

Turbine meters are used in applications with flow rates from fractions of gallons per minute to tens of thousands of gallons per minute, with pressures to 7500 psi, and with temperatures from $-400°F$ to $1000°F$. They can be used with any reasonably clean fluid and have a rangeability of 20 to 1.

The output relationship is linear within the particular transducer's limits for flow rate and viscosity. A linearity of $\pm 0.5\%$ of flow rate and a repeatability of 0.02% of rate can be achieved using this technique.

Bearing friction, fluid and magnetic drag, and swirl in the fluid stream may cause errors in these transducers. These errors should be minimized in the mechanical design. Swirl effects can be greatly reduced by the use of flow straighteners. The magnetic drag is not significant except in miniature transducers. Thrust bearing friction can be reduced using the hydrodynamic forces to balance the axial loads on the rotor.

The transducer case must be made of a nonmagnetic metal and it is usually threaded for the detector coil. Since the amplitude of the output voltage from the coil is a function of the distance between the pole piece and turbine blade tips, the threaded coil can be used to adjust this voltage (Fig. 4-6).

Most turbine flowmeters use an electromagnetic coil of the permanent magnet type. The turbine blades are then made of a ferroelectric material. An accuracy of 0.5% can be achieved in the linear portion of the flow coefficient curve. However, along with its advantages a turbine meter installation complete with pickup electronics can easily be more expensive than similar head meter installation.

Other disadvantages include

1. Abrasive materials can quickly wear out the internal bearings.
2. Pulsating flows or water hammer can cause damage to the moving parts.
3. High-viscosity liquids can affect measurements.

The advantages of turbine meters can be summarized as

1. High-pressure, high-flow measurement capabilities over a wide temperature range.

Figure 4-6. The case of many turbine flowmeters is threaded for the detector coil to allow adjustment of the output voltage. (*Courtesy AMETEK, Inc.*)

2. High accuracy and repeatability.
3. Good flow range.
4. Good response times to flow changes.
5. Short piping approaches required and flange ends allow quick replacement (Fig. 4-7).
6. They can be converted to mass flow with compensating hardware or software.

VARIABLE-AREA METERS

Variable-area meters may use a float in a tapered section of tubing (these are called rotameters), a spring-restrained plug, or a spring-restrained vane. A displacement of these elements causes the area of the flow passage to vary while the differential pressure or head remains constant. The displacement is then measured to provide an output proportional to the flow rate.

Rotating elements include spring-restrained turbines installed in pipe sections and spring-restrained propellers installed in flow streams. These turn at an angular speed proportional to the flow rate and are displaced along the axis of rotation.

Rotameters measure the stream flowing through a tapered vertical tube using a float as shown in Figure 4-8. The float will rise in response to the rate of flow and will stabilize for constant flow.

78 TRANSDUCERS FOR AUTOMATION

Figure 4-7. Turbine meters with flange ends allow quick replacement in process systems. (*Courtesy AMETEK, Inc.*)

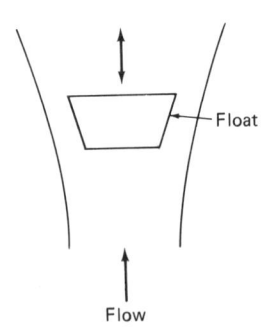

Figure 4-8. Rotameter (variable-area) flow sensor.

The density and viscosity of the liquid or gas being measured by the rotameter can affect the float position. The area around the float increases as the float rises in proportion to the flow rate. If the rotameter tube is glass, the flow may be read by optical methods. With opaque tubes, magnetic techniques can be used to monitor the float position. The rangeability is 12 to 1 and an accuracy of 1% is possible. A number of sizes allow capacities from 0.5 cm^3 to 5000 gpm. Temperatures to 1000°F and pressures to 2500 psi can be accommodated. Figure 4-9 illustrates some typical applications.

Rotameters are characterized by a low pressure drop. They allow some solids in the fluids and provide an average reading for pulsating flows. Their main

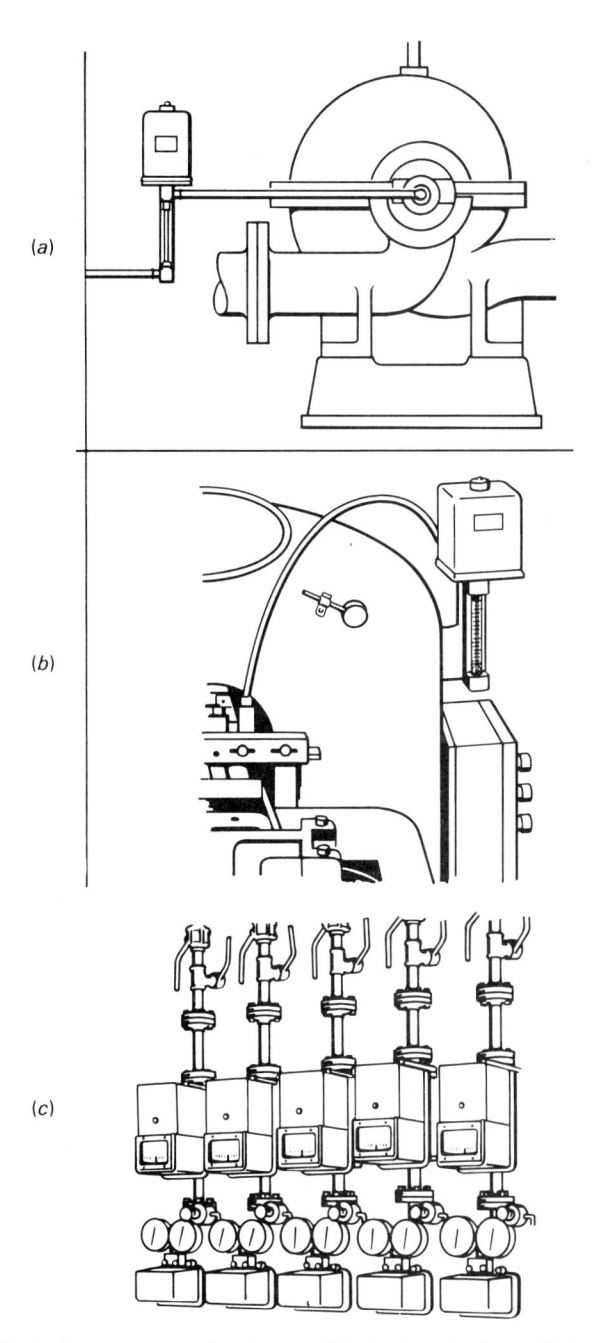

Figure 4-9. Typical rotameter applications. *a*. Monitoring the flow of lube oil for pump bearings. Used in a soldering operation, unit *b* monitors the flow of flux. *c*. Controlling the rate of flow of a binder in the manufacture of insulation. (*Courtesy AMETEK, Inc.*)

80 TRANSDUCERS FOR AUTOMATION

drawback is the required vertical installation along with a high cost for sizes larger than 2 in.

The flow-induced deflections of vanes and supports are sensed in some transducers by displacement sensors or strain gauge bridges in several rotameter designs.

FLUID CHARACTERISTIC SENSORS

A number of flow sensors use the characteristics of the moving fluid to measure flow. One technique is to use a heated wire as a hot-wire anemometer sensor. The heated wire will tend to transfer more heat as the velocity of the surrounding fluid increases. The resultant cooling effect causes the resistance of the wire to decrease.

Another type of sensor uses a fluid containing a small amount of radioactivity. One then monitors the change in ionization current to determine the flow velocity.

If a slightly conductive fluid flows through a transverse magnetic field, it will provide an increasing voltage for an increasing flow velocity. In another type of flow sensor, the boundary layer of a moving fluid is heated by a small heating element and the convective heat transfer, as measured by a temperature sensor located downstream from the heater, changes with a change in the flow velocity.

Those transducers which convert flow rate into a change of resistance all make use of the fluid characteristics. The methods used to obtain the resistive element response can be grouped into three sensor types:

1. Hot-wire anemometers which are used primarily for air flow measurements.
2. Thermal and boundary layer flowmeters.
3. Oscillating-vortex flowmeters.

Hot-wire anemometers use a wire element which is located normal to the flowing stream (Fig. 4-10). The wire is heated electrically and cooling due to

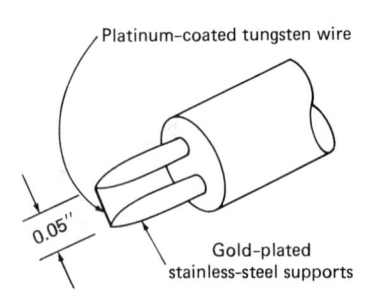

Figure 4-10. Typical hot-wire sensor.

the flow changes affects the resistance of the wire. The resistance changes are calibrated to indicate the flow velocity.

Some hot-wire anemometers maintain the wire temperature constant and measure the current required to maintain this temperature. This measured current is then a calibrated indication of the flow. Others use a nonlinear amplifier to maintain a constant current through one side of a bridge circuit as shown in Figure 4-11.

Hot-wire anemometers can be used to measure mass flow provided that the product of the thermal conductivity, specific heat, and density remains constant. This is true for many gases at low pressure.

The hot-wire anemometer has the greatest application in the measurement of low flow rates of gases. It is also used to determine gas velocity. Anemometers offer a good sensitivity to flow changes but complete installation can be costly.

Flow rate can be measured using heat transfer methods with one of two types of instruments: the thermal flowmeter or the boundary layer flowmeter. Both of these use an electrical heater to increase the heat of a portion of the moving fluid as shown in Figure 4-12.

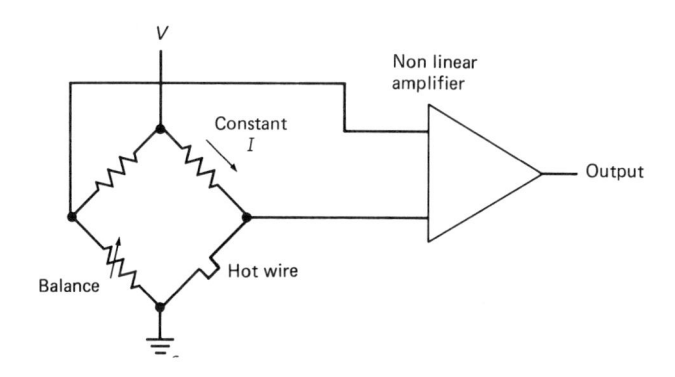

Figure 4-11. Constant-current anemometer system.

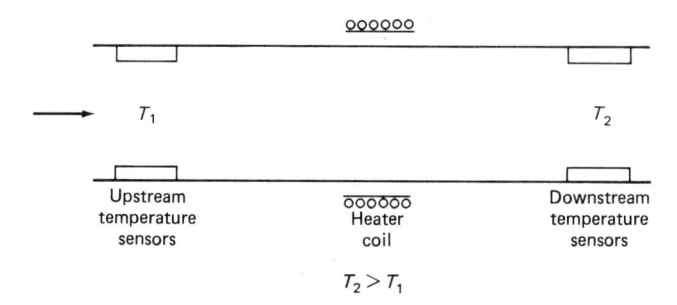

Figure 4-12. Heat-transfer technique for flow measurement.

82 TRANSDUCERS FOR AUTOMATION

Temperature sensors, which may be resistive but more usually are thermoelectric, then measure the temperature of the fluid upstream and downstream from the heater. The heat that is transferred between the heater and the downstream sensor is found either by measuring the temperature differential between the two sensor areas at a constant heat input or by measuring the change in heater current to keep this temperature differential constant.

Early thermal flowmeters used a heater immersed in the fluid. Since the entire core of the fluid was heated, the heater power required for some of the larger units was in the kilowatts. The high power requirement was a limiting factor in the application of these early thermal flowmeters.

The power is greatly reduced with the boundary layer flowmeter in which only the thin boundary layer of the fluid adjacent to the pipe wall is heated. The temperature sensors are mounted flush with the inside pipe surface, and the heater surrounds the outside of the pipe or is embedded in the wall of the pipe.

The heat input varies with the mass flow rate and the temperature differential. The relation between these quantities is more complex than for the internally heated thermal flowmeter. The heater power requirements are usually less than 50 W for boundary layer units.

Another type of thermal flowmeter uses a constant-ratio bypass which samples the gas flow in the main duct. Although this is not a boundary layer device, it has a relatively low power consumption. Power requirements for the transducer and readout are 15 W at 115 V ac; about 5 W is used by the transducer itself.

The forced-vortex meter, which is also called the precessing-vortex meter or swirl meter, uses the vortex precession of the fluid. The principle of vortex precession is based on the fact that when a rotating body of fluid enters an enlargement as shown in Figure 4-13 under the proper conditions, the flowing fluid will oscillate with a well-defined frequency that is proportional to flow.

If we detect these oscillations, the output will be a train of pulses with a frequency that is proportional to the flow rate. This train of pulses is a function of the fluid generating a hydrodynamic precession of the flow.

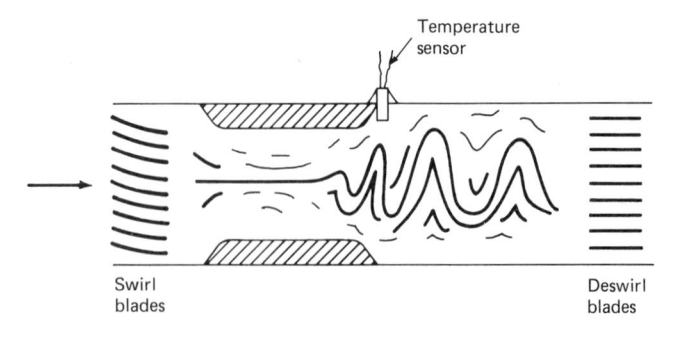

Figure 4-13. Forced-vortex flowmeter.

Swirling can be imparted to the fluid by a set of blades. Downstream of these blades is a Venturi-shaped contraction and expansion of the flow passage. In the region where the cross section enlarges, the swirling flow precesses as it leaves the axial path of the meter center line and takes on a helical path. The frequency of this precession is a function of the flow rate.

The oscillations of the fluid create variations in fluid temperature which can be detected by either a platinum film sensor or a thermistor-type resistive temperature sensor. Deswirl blades are used to straighten the flow as it leaves the flowmeter. This type of flowmeter has a maximum linear range of 100:1, within ±1% of rate for an output frequency of 10 Hz to 2 kHz.

The vortex shedding flowmeter is of a similar design, but it makes use of hydrodynamic principles. A nonstreamlined obstruction is placed in the pipe and the fluid does not flow by it smoothly. Eddies or vortices are formed which grow larger, and eventually detach themselves from the obstruction (Fig. 4-14). The detachment is known as shedding. It occurs alternately at each side of the obstruction. The vortices form trails downstream of the obstruction. The shedding frequency of the vortices is directly proportional to the flow rate.

In the vortex shedding flowmeter the obstruction acts like an orifice plate, and the output signal is like that of a turbine-type flowmeter.

The vortex shedding flowmeter senses flow velocity; thus it acts as a volumetric measuring device. It can be used as a mass flowmeter if the fluid density is known. Advantages include a wide range, universal calibration, and no moving parts. However, the vortex shedding flowmeter cannot be used in laminar flow and the Reynolds number of the fluid must be at least 3000. Vortex shedding occurs in both liquid and gaseous fluids.

Another flow sensing technique uses radioisotopes and detectors for the nuclear radiation that is imparted to the fluid. A source is mounted on the outside of the pipe upstream from the detector (Fig. 4-15).* Neutrons from the source

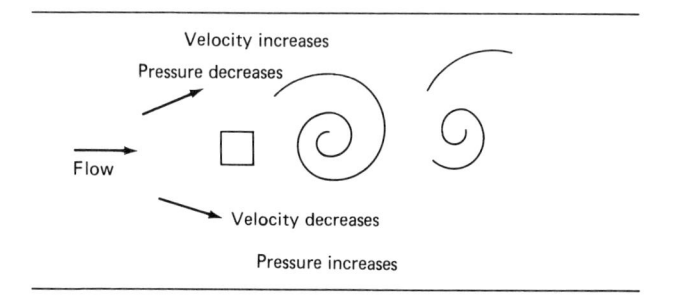

Figure 4-14. Vortex shedding.

---

*Radiation sources and safety considerations are discussed in detail in Chapters 5 and 7.

84 TRANSDUCERS FOR AUTOMATION

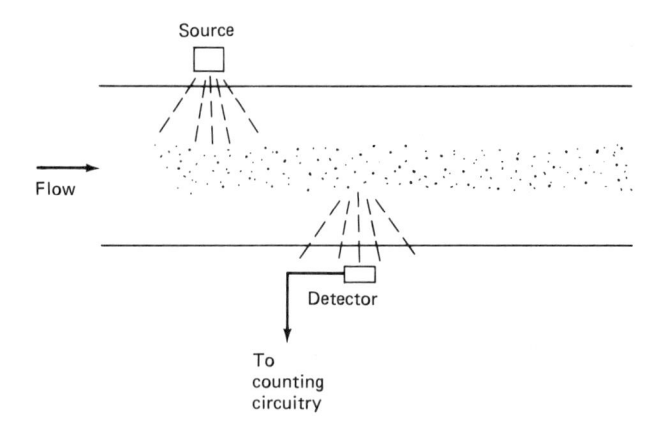

Figure 4-15. Nucleonic flow measurement.

collide with the moving fluid and cause particle and electromagnetic radiation to be emitted. Most of this emitted radiation occurs at the location of the source, but some is emitted as the fluid passes by the detector. The detected counts of radiation are a function of the flow rate of the fluid. Another nuclear technique obtains a count by adding small amounts of a radioactive trace element to the fluid.

Nucleonic flowmeters offer no obstructions to the fluid path. They can be useful in the measurement of difficult fluids such as multiphase variable-composition fluids, slurries, and suspensions.

ELECTROMAGNETIC FLOWMETERS

Electromagnetic flowmeters employ Faraday's law, which states that a relative motion, at right angles between a conductor and a magnetic field, will induce a voltage in a conductor. This voltage is a function of the relative velocity of the conductor and the intensity of the magnetic field.

An electromagnetic flowmeter as shown in Figure 4-16 is made from non-magnetic materials and uses a conductive liquid. Magnetic coils provide the magnetic field, and as the liquid moves through this field a voltage is generated proportional to the fluid flow rate.

Electromagnetic flowmeters are calibrated to detect the liquid velocity, so the entire cross-sectional area of the pipe must be full and there should be no gas bubbles in the liquid since these will also cause errors.

Electromagnetic flowmeters have an output that ranges from microvolts to millivolts. Proper installation and grounding are important for accurate flow sensing.

This type of flowmeter is not affected by changes in liquid density or viscosity. Turbulence of the liquid and variations in piping have only limited effects. The

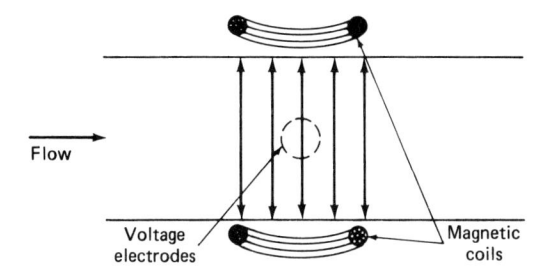

Figure 4-16. Electromagnetic flowmeter.

main requirement is that the conductivity of the fluid be greater than about 10^{-8} mho/cm^3. Mixed-phase fluid may cause conductivity variations and gross measurement errors. Direct-current coil excitation can cause electrolysis problems and is not used. The linearity of output voltage with flow rate can be an advantage in many process automation applications.

SONIC AND ULTRASONIC FLOWMETERS

Piezoelectric flowmeters have been used to measure fluid flows since the 1950s. The basic configuration has two transducer pairs which establish upstream and downstream sonic paths diagonally across the fluid as shown in Figure 4-17. The difference in propagation velocity (the Doppler effect) between the two paths is calibrated to measure the flow rate. A flow section is normally used to secure the transducers.

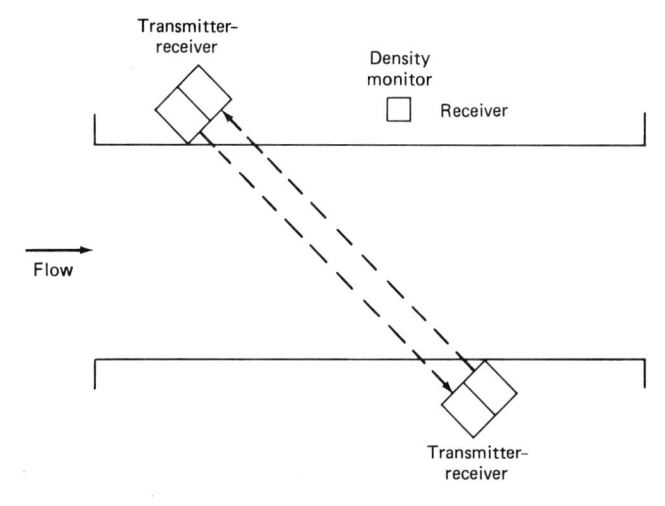

Figure 4-17. Doppler ultrasonic flowmeter.

Sonic and ultrasonic flowmeters may be used with liquids or sonically conductive slurries through either closed pipes or open channels. They are useful in water and waste plant automation as well as process automation. They can be used to monitor bulk material flow on automatic conveyors or chutes and other material handling machines.

Ultrasonic systems may use level measurement, Doppler, or backscatter Doppler. The one common component is the sensor itself, which is usually the most critical part of the system.

In a level measurement system, flow is monitored by measuring the height of liquid flowing in an open flume using the sound transmission of air. These transducers operate in the low ultrasonic portion of the spectrum.

Doppler systems operate at higher frequency with devices operating in the range of 100 kHz to 1500 kHz. In some cases the transducers are coupled to the wall of the conduit through which the fluid flows. The ultrasonic energy is transmitted through the wall and the fluid at an oblique angle to the flow. A second transducer is placed on the opposite side of the stream to receive the signal. A third transducer is used in some systems to monitor any density variations that may occur as shown in Figure 4-17.

WEIRS AND FLUMES

Weirs and flumes are open-channel devices which are used to develop a liquid head for measuring flow rate. They are often used for water flow, and can be used wherever the flow is in open channels or in pipes and conduits that are not completely filled with liquid. The conditions must provide a tranquil surface for the head measurement and the liquid approach must be a straight run.

A weir is basically a dam with a notched opening in the top through which the liquid flows (Fig. 4-18). It is a simple method and it can be relatively accurate for measuring the flow of liquids under the proper conditions.

The flow rate (FR) is determined by measuring the head above the lowest point of the weir opening through which the liquid flows. This height may be measured by a float installed in a box, called a stilling well, which is a part of the total weir installation. The float is located here so that it is not disturbed by the velocity of the flow, or by any turbulence in the stream.

The rectangular notch weir is the oldest type and tends to be the most commonly used because of its ease of construction. The trapezoid notch, which is also called the Cippoletti, is designed so that the trapezoidal sides produce a flow correction to allow the flow to be proportional to the length of the weir crest.

The V notch offers the widest range for a single size, since the small opening in the bottom of the V can accommodate small flows and the upper portion larger flows. It has the greatest head loss because of its shape.

$$FR = 3.33\,(L - 0.2h)h^{1.5}$$

a

$$FR = 3.36\,L h^{1.5}$$

b

$$FR = 2.48\,\tan\tfrac{1}{2}\theta\, h^{2.5}$$

c

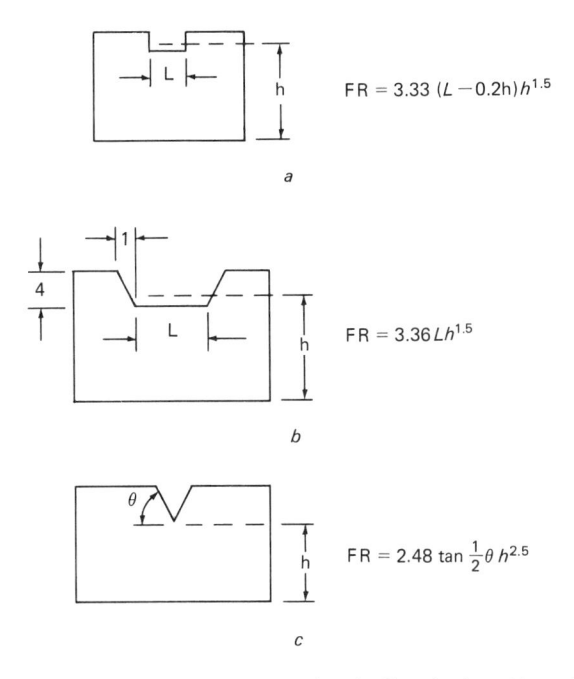

Figure 4-18. Weirs: *a*. rectangular, *b*. Cippoletti, *c*. V notch.

A weir must be cleaned periodically if the liquid being measured contains entrained material since deposited materials will produce errors in the flow rate.

Parshall flumes (Fig. 4-19) are self-cleaning and operate with a small loss of head. The head loss is about 25–30% that of weirs. Parshall flumes may be used where sand, grit, or other heavy solids are present in the fluid stream. They should be used for streams in which the flow velocities are only moderate. They can be affected by bends or other objects that can cause eddies, waves, or uneven flow patterns.

The flow rate through weirs and flumes measured by the level or head gives a direct relation to the volume flow. No correction is required for the liquid density. Errors of 2% are possible and flow rates from a few gallons per minute to millions of gallons per minute can be measured using these devices.

In the rectangular and trapezoidal weirs and the Parshall flumes, flow is proportional to the three-halves power of the measured head. In V-notch weirs, the flow is proportional to the five-halves power of the head. The actual relationships are shown in Figure 4-18. With the proper installation these units can provide a wide range of flow measurements. It is possible to achieve a flow range of 20 to 1 for rectangular and trapezoidal weirs and Parshall flumes, and up to 40 to 1 for V-notch weirs.

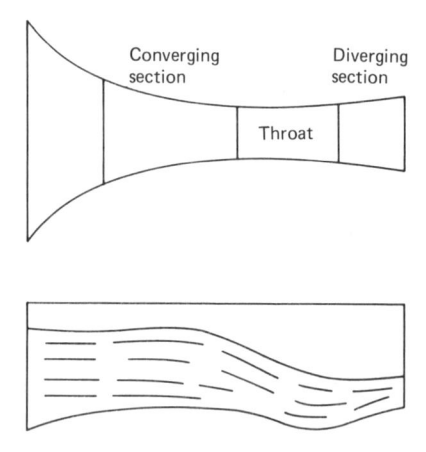

Figure 4-19. Parshall flume.

POSITIVE DISPLACEMENT METERS

Positive displacement meters divide the flow into known volumes and determine the flow by counting the number of volumes or revolutions. These are mechanical meters with moving parts that physically separate the fluid into increments.

The energy used to drive these parts comes from the flow stream and produces a pressure loss between the inlet and the outlet. The accuracy is dependent upon the clearances between the moving and stationary parts and the meter accuracy tends to increase for the larger sizes. Positive displacement meters with the proper tolerances can be within $\pm 1\%$ total error over flow ranges to 20:1. They are useful in batch processing and mixing or blending automation.

Since positive displacement meters need small clearances to keep their accuracy, the measured liquids must be clean.

Positive displacement meters that do not require electrical power or an air supply may be damaged by exceeding their capacity. Some types use electrical drive motors to deliver a given volume, and electric pneumatics or hydraulic power may be used for control purposes.

Piston pumps act as metering pumps to inject an exact amount of fluid into a flow line or a collecting vessel. The piston pump is generally the reciprocating type where a piston or plunger delivers a fixed volume on each stroke. In the basic single-action unit the fluid is drawn into the piston cavity through an inlet valve as the piston is moved back. The metered fluid is discharged into the flow line or a vessel as the piston moves forward. As the operation repeats itself, the pump delivers a fixed volume using this pulsating flow.

Piston meters are useful in delivering controlling volumes at high pressures. The pulsating flow can be averaged out using reservoirs. The amount of flow can be changed by varying the length of the piston stroke or by changing the pumping speed. The stroke adjustment can be varied manually or automatically depending on the construction of the pump.

We have considered a number of flow sensing elements which may be used for the measurement of volumetric flow rate. Mass flow rate can be determined from this measurement and a simultaneous measurement of density. A microcomputer could use the density and volumetric flow rate inputs to provide the mass flow rate. A number of sensing elements can be used that have an output which is directly proportional to the mass flow rate. These transducers are considered below.

MASS FLOWMETERS

The transducers that are available for the direct measurement of mass flow avoid the need for separate density and volumetric flow rate measurements. To compute mass flow rate the twin-turbine mass flowmeter uses two bladed rotors with different blade angles. The rotors are coupled by a spring to allow a relative angular motion between them. As a result of the blade angle difference, the two turbines attempt to rotate at different speeds but are restrained by the spring coupling. They assume an angular displacement with respect to each other which is proportional to the flow momentum. If the two rotors are considered as a unit, they function as a volumetric flowmeter. The turbine speed is a function of the average flow velocity. The angular displacement is sensed as the phase angle difference using the outputs of the two transduction coils.

The time that it takes for the phase angle to change can be used to measure the mass flow rate. Digital counters implemented in either hardware or software can be used to interpret the dual output.

Another type of twin-rotor mass flowmeter utilizes the angular momentum of mass flow. This design uses two similar axial flow rotors. An upstream rotor is driven at a constant low speed by a synchronous motor. This rotor is used to provide a constant angular momentum to the fluid. A downstream rotor is used to remove all angular momentum from the fluid. As a result, a torque is exerted in accordance with Newton's Second Law. The torque that is produced is linearly proportional to the mass flow rate. The angular displacement that results is proportional to the torque from a restraining spring. A reluctive transducer is used to convert the angular displacement into an ac output voltage.

The gyroscopic mass flowmeter operates on the gyroscopic vector relationship. As shown in Figure 4-20, the liquid is forced to flow around a loop of pipe which is in a plane perpendicular to the input line. After one cycle around the loop, the fluid is returned to the input axis. During this rotation, the fluid develops an angular momentum similar to the rotor of a gyroscope.

If the loop is vibrated through a small angle about an axis in the plane of the loop, the vibration produces an alternating gyrocoupled torque about the or-

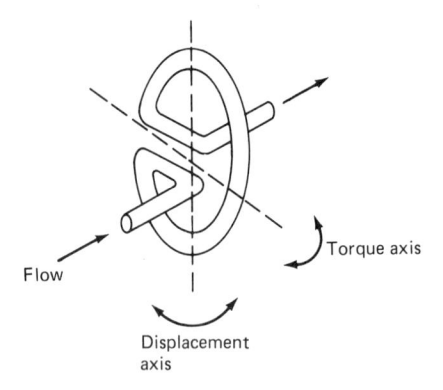

Figure 4-20. Gyroscopic mass flowmeter.

thogonal axis. The peak amplitude of this torque is proportional to the mass flow.

The torque is forced to act against restraints, which produces an angular displacement about the torque axis. Sensors then convert the displacement to an electrical signal. The peak signal is a measure of the mass flow rate.

Flow monitoring applications in factory automation will continue to expand because of the increased emphasis on energy conservation. Many manufacturers have drastically lowered energy consumption with improved instrumentation.

New applications of physical principles, such as hydrodynamic oscillation, are constantly being applied to flowmeters in order to increase reliability, extend applications, and reduce costs.

One design uses a self-excited oscillating prism to produce a signal over the range from 4 to 200 Hz. The output is linear with flow and is relatively unaffected by fouling within the meter.

A prism-shaped oscillator, with a pendulum support, is excited by the flowing fluid. The frequency of the oscillation is linearly proportional to the flow. Magnetic induction is used to detect the vibration and converts it into an electrical signal in one of three different ways: (1) by a frequency converter into an analog current proportional to the flow, (2) into a digital signal by counting the number of oscillations in a given time period, or (3) into a count representing the total amount of fluid that has passed through the pipe.

In another, newer type of flowmeter the meter body acts as a fluidic oscillator and produces a frequency that is linear with the volumetric flow rate. The oscillations are detected and amplified to provide a digital pulse output.

The meter body in this flowmeter is shaped so that the flowing stream attaches itself to one of the side walls using the Coanda effect. A portion of the flow is diverted back through a feedback passage to a control port. This feedback acts upon the main flow and diverts the main flow to the opposite wall where the feedback action is repeated. The result is continuous self-induced oscillations with an oscillation frequency that is a linear function of the flow velocity.

Accurate measurements under conditions that are also compatible with the automation system are always important. While there are flowmeters for all sorts of situations for measurement under almost any condition, most of these produce an analog output that must be converted to a digital format in order to interface the meter to a digital bus. Some flowmeters such as those discussed above as well as the turbine and vortex types can produce a digital frequency output.

EXERCISES

1. Compare the operation and construction as well as the advantages and disadvantages of the Venturi tube and orifice plate.
2. List the merits and limitations of the industrial type of Pitot tube.
3. Discuss how the Venturi tube can be used for liquid flow measurement with a monitoring system for a local microcomputer alarm control system located within 30 ft of the pressure duct.
4. Water is flowing down a vertical pipe 8 ft long. The pipe has a diameter of 2 in. If the quantity of water is 40 gpm nominal with expected variations of ± 30, describe a flow monitoring and pumping control system using a low-cost flow sensor. Select an alternative transducer as well.
5. Describe the operation of a mass flowmeter that uses gyroscopic motion.
6. Discuss the importance of a temperature sensor with a short time constant in the oscillating-vortex flowmeter. Explain what methods you could use to determine the average velocity of air flow through a large circular pipe and a large rectangular air duct.
7. Describe the operations required to measure liquid flow in an open channel by means of a simple microcomputer system. Show the operations with a flow chart.
8. What is the function of the downstream rotor in a twin-turbine flowmeter?
9. Costs are one of the major items to be considered in selecting a flow sensing unit. Select several types of flowmeters and discuss how line size as a measure of capacity can form a basis for making cost comparisons.
10. The various head meters possess some unique features which must be compared whenever one is faced with a decision as to the type to be installed. Discuss these features as they relate to the application criteria. Summarize your results in a head meter selection chart.
11. Discuss how each of the flow sensors listed below might be used to provide flow information for a flow that is mainly liquid, with some undissolved solid matter; the flow is fairly rapid and in a 3-in.-diameter plastic pipe: (a) electromagnetic, (b) turbine, (c) heat transfer, (d) ultrasonic, (e) vortex shedding.
12. Discuss the use of each flow sensor listed below for a flow of dilute sulfuric acid in a stainless-steel pipe of 2.5 in. diameter; the flow is medium speed with no turbulence: (a) electromagnetic, (b) turbine, (c) ultrasonic, (d) vortex shedding.

13. Select a flowmeter to be used for the following process automation applications; discuss your answers: (a) gaseous flow; (b) homogeneous nonconductive fluid flow; (c) pure, nonionized water; (d) conductive fluid flows; (e) flows of small-particle solid materials, slightly acidic.
14. Discuss the use of either an ultrasonic or turbine flowmeter for the following applications: (a) medium flows of conductive fluid; (b) very low flow rates; (c) heat velocity of nonconductive fluid; (d) the output must be a sine wave, with constant amplitude and variable frequency; (e) the output must be a pulse train, of variable pulse repetition rate.
15. Mass flow can be inferred from measurements made with an orifice plate or any of the head meters with additional equipment used to measure, compensate, and make computations for specific gravity, flow temperature, and pressure. Show how this can be done with a low-cost microcomputer. Draw a flow chart for the computations.
16. There are several ways of obtaining mass flow rate from conventional weighing devices. Weighing belts can be provided with a belt speed sensor; the signal from this sensor is combined with the weight signal to provide a mass flow rate signal. Show how this can be accomplished with a low-cost microcomputer. Draw a flow chart for the computations.

5
DENSITY AND WEIGHT MEASUREMENT

DENSITY

The density of a substance is one of its basic characteristics, since it can be used to provide information concerning composition, concentration, mass flow, and, in fuels, caloric content. A measurement of specific gravity relative to air can provide information on stream compositions.

A density measurement at actual operating conditions can often eliminate the need for separate pressure, temperature, supercompressibility, or humidity measurements if the purpose is to determine the mass flow. Consider also that all mass flowmeters can be used as densitometers, provided that the volumetric flow rate is constant.

The density of solids is usually measured by weighing the solids in air and then weighing them in a liquid of known density. The density is

$$D = \frac{W_a D_l - W_l D_a}{W_a - W_l},$$

where

D_a = density of air,
D_l = density of liquid,
W_a = weight in air,
W_l = weight in liquid.

Density is defined as the quantity of matter per unit volume. The most common unit is grams per cubic centimeter but pounds per cubic foot are also used.

Specific gravity is defined as the ratio between the density of a process material and that of water or air at specified conditions. This ratio has no true units associated with it; however, various scales divided into units are commonly used to describe the specific gravities of different materials (see Table 5.1).

Both density and specific gravity characterize the same physical property of the process media and they must be defined at some specified temperature. In the case of specific gravity, the temperatures might be different for the process and the reference fluid. For example, a specific gravity table might list a process

Table 5.1. Commonly used specific gravity units.

Scale in API hydrometer degrees for petroleum products:

$$\frac{141.5}{\text{S.G. @ 60°F}} - 131.5 = °\text{API}.$$

Scale in Balling degrees (°Ba) is used in the brewing industry and indicates weight percentage of dissolved solids in 60°F water.

Scale in Barkometer degrees is used in the tanning industry. Each degree is 1/1000 of an S.G. unit above or below 1.000 [+10°Bk means S.G. = 1.010 or (S.G. − 1.000) × 1000 = °Bk].

Scale in Baumé degrees for light liquids including ammonia:

$$\frac{140}{\text{S.G. @ 60°F}} - 130 = °\text{Be}.$$

Scale in Baumé degrees for heavy liquids including acids:

$$145 - \frac{145}{\text{S.G. @ 60°F}} = °\text{Be}.$$

Scale in Brix degrees (°Br) is used in the sugar industry and represents percent sugar by weight in 60°F water solution.

Scale in Quevenne degrees is used in the milk industry. Each degree is 1/1000 of an S.G. unit above 1.000 [40°Q means S.G. = 1.040 or (S.G. − 1.000) × 1000 = °Q].

Scale in Sikes, Richter, or Trailers degrees gives the volume percentage of ethyl alcohol in water. Proof stands for twice the volumetric percentage of ethyl alcohol in water.

Scale in Twaddell degrees is applied to fluids heavier than water. Each degrees is 5/1000 of an S.G. unit above 1.000 [10°Tw means SG = 1.050 or (SG − 1.000) × 200 = °Tw].

---

fluid as having a $0.87_{80/40}$ specific gravity. This liquid at 80°F (27°C) would have a density 0.87 times that of water at 40°F (4.4°C).

For gases, the specific gravity is normally based on standard conditions. Thus, both the process vapors and the reference air density are measured at 60°F (16°C) and atmospheric pressures. For perfect gases, the ratio of molecular weights gives the same data.

Density may be detected indirectly through the measurement of another process property. The measurement of boiling-point elevation is one common example. Resistance elements can be used to compare the temperature of the boiling process sample with that of boiling water at the same pressure. The differential temperature scale for a particular solution can be calibrated in terms of density.

GAS DENSITY DETECTORS

Since gases are compressible materials, the same detector cannot usually detect both the specific gravity (or molecular weight) and the density at operating conditions. Thus, two types of gas sensors have evolved: (1) those that determine gas density for the stream composition and (2) those that calculate the mass flow rate from orifice or volumetric flow data. The first type uses a comparison of sample gas density to air at ambient conditions while the second requires the measurement of process gas density at operating conditions.

The stream composition units are calibrated either in specific gravity units based on air, or in molecular weight units. The scales are interchangeable since

$$SG = \frac{\text{Molecular weight in gas}}{\text{Molecular weight in air}}$$

Density sensors for measuring the gas at operating conditions can also be used to determine the compressibility factor of the vapor. Supercompressibility is a measure of the deviation of actual from ideal gas behavior and is defined as the ratio between the actual specific weight and that based on the ideal gas laws. If the gas pressure, temperature, specific gravity, and flowing density are measured, then the supercompressibility factor becomes

$$Z = \frac{P\,SG}{\rho T R_a}$$

where

Z = supercompressibility factor (dimensionless),
P = operating pressure (lbf/in.^{2}),
SG = specific gravity (dimensionless),
ρ = flowing density (lbm/ft^3),
T = absolute temperature (K),
R_a = gas constant for air (53.3 ft-lbf/lbm K).

The important difference between these techniques is that for the mass flow rate determination there is no need for separate pressure, temperature, supercompressibility, or humidity measurements. When gas composition is important these variables must be taken into consideration and their effects compensated for.

ANGULAR POSITION LIQUID DENSITY SENSORS

In the angular position liquid density sensors, the process fluid sample flows continuously through the detector (Fig. 5-1) at less than 30 gph (114 L/h). The

96 TRANSDUCERS FOR AUTOMATION

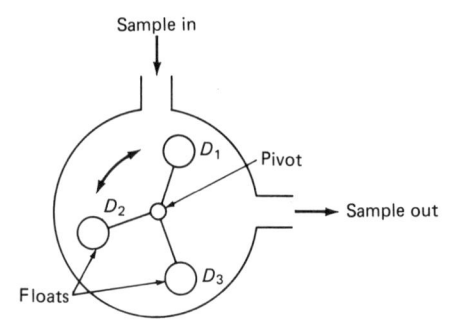

Figure 5-1. Angular position density sensor.

gauge chamber contains three displacer floats, each with a different density and volume. The floats are spaced 90°–100° apart and connected to a common shaft.

As each fluid density sample is taken, it positions the shaft at some angular position which is a function of the buoyant force. The angular position of the assembly is transmitted to the electrical components through a magnetic coupling.

Since the floats are made with solid materials, process pressure changes have no effect on their volume. The effects of process temperature variations is also small because the float materials have a low coefficient of expansion. Changes of 200°F (111°C) produce a measurement shift of only 0.0001 times the specific gravity. Detection accuracy is ±0.0005 to 0.0001 SG and is unaffected by the span.

The sample flow rate should be constant and not contain any particles in excess of 20 μm. These devices are limited to clean, low-viscosity services, for viscosities up to 500 cP (0.5 Pa s). Temperature compensation can be obtained with the use of additional sensing and computing elements.

BALL LIQUID DENSITY SENSOR

The ball-type density meter uses a number of $\frac{1}{4}$-in. (6.3-mm) diameter hollow opaque silica glass balls. Typically 10 to 50 balls are used. Silica glass has a low thermal expansion of 0.5 parts in 10 per degree Fahrenheit. The balls are free to move in tubes that are immersed in the fluid to be measured. Each ball has a different density; for example, in the range of 0.7 to 0.8 specific gravity, 10 balls are graded as 0.70, 0.71, 0.72, . . . , 0.79 specific gravity.

Fiberglass probes conduct light through one side of the tubes containing the balls. Probes through the other side of the tubes sense the light if the ball is not intervening. The last ball to float determines the density of the fluid. Photodiodes are used to produce electrical signals from the light signals to enter a data bus or provide a readout. The tubes may be located near the bottom of the tank and are twice as long as the ball train which they enclose. The tubes may also extend to the full height of the tank to detect variations and density stratifications. The fluid height is indicated by the ball of least density. The spherical shape of the floats reduces friction in the tube walls, allowing this type of density meter to be mounted at an angle of up to 85° to vertical.

An advantage of this type of density meter is its intrinsic safety. It can be used in explosive mixtures without the danger of causing an explosion. It is also not affected by most electric or magnetic fields.

CAPACITANCE LIQUID DENSITY METER

The capacitance type of liquid density meter uses two concentric cylinders into which the medium to be measured is sampled. A bridge circuit is used to measure the capacitance between the cylinders. The capacitance is proportional to the dielectric constant, which is, in turn, proportional to density.

This method can also be applied to measure the total fluid in a tank in pounds. The probe, made up of the two concentric cylinders, extends from the top to the bottom of the tank and fills to the level in the tank. The product of the density, found from the dielectric constant, and the volume is measured and displayed as the number of pounds in the tank. For irregularly shaped tanks, the probes may be contoured to match the shape of the tank, providing a linear output. Several probes can be connected in parallel to sum the mass in several tanks or to compensate for tilting of the tank, which is usual in aircraft applications.

DISPLACEMENT LIQUID DENSITY SENSOR

Constant-volume, variable-weight density elements are used in the displacement-type liquid density sensors, which represent one of the basic methods of weighing a fixed volume. The float in these sensors is submerged in the process liquid, displacing a fixed volume. The float buoyancy is a function of the liquid density, and by sensing the buoyant force, the fluid density can be measured. Three types of displacement density element will be considered: (1) the calibrated-chain-balanced sensor, (2) the electromagnetic-force-balanced sensor, (3) the conventional displacer float sensor.

The chain-balanced float is based on the hydrometer concept shown in Fig. 5-2. In this type of density indicator a submerged float and chain assembly displaces a fixed fluid volume. The float buoyancy is a function of the liquid density and an increase in density causes the float to rise. As it rises, it supports a greater portion of the calibrated chain.

The density-related vertical position of the float can be detected by an in-

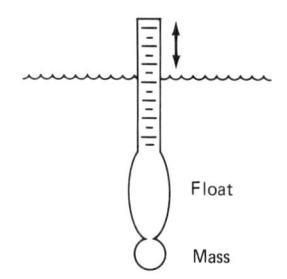

Figure 5-2. Hydrometer concept.

ductance pickup, which consists of a magnetic core inside the float and a three-winding differential transformer. The float movement causes an inductance change with a resultant change in the differential voltage output.

The wetted parts of this unit are available in a variety of corrosion-resistant materials including Pyrex and clear plastics for the chamber. Specific gravity spans from 0.01 to 0.5 can be obtained within the limits of 0.5 and 3.5 SG. The inaccuracy is about 1–3% of full span and the detector can be exposed to pressures up to 125 psig (861 kPa) and temperatures to 400°F (204°C).

The sample flow rate through the instrument must be between 5 and 30 gph (19 and 114 L/h) for process streams having less than 50 cP (0.05 Pa s) viscosities. Higher viscosities or flow rates introduce velocity and friction effects which reduce the accuracy.

The chamber can also be furnished in corrosion-resistant metallic or plastic-lined materials, and the pressure and temperature limits are increased to 500 psig (3.4 MPa) and 500°F (260°C). The speed of response is a function of sample flow rate, and averages about 30 s.

When the process temperature varies, it is necessary to use automatic compensation, which corrects the apparent fluid density to a predetermined base temperature. A resistance-type temperature sensor is used, and the instrument includes adjustments to permit selection of the compensation values.

DISPLACER FLOAT DENSITY SENSOR

Displacer-type float density sensors are all based on the displacement-type level devices considered in Chapter 7, but some features are peculiar to the application of density detection. The basic difference between a level and a density displacer sensor is that the displacer float in level sensing is only partly covered by the process liquid, while for density detection it is totally immersed, so any changes in buoyant force appear as density variations.

The major limitations of displacement-type density sensors relate to the types of process fluids they can handle. Velocity effects can be eliminated by controlling the sample flow rate and by the use of piezometer rings. Viscous or slurry process streams can cause material buildup on the float, which will change its weight and affect the measurement.

Displacement-type density sensors should only be used for clean, nonviscous process streams with sample flow rates in the low gallons or liters per hour range.

ELECTROMAGNETIC SUSPENSION DENSITY SENSOR

Electromagnetic suspension density sensors employ a float which is suspended electromagnetically and totally immersed in the process fluid. The position of the plated ferrous alloy float is detected by a pair of sensing coils as shown in Figure 5-3.

The float must be slightly more dense than the maximum density of the fluid to be measured so that it will sink. However, it is prevented from doing so by the solenoid directly above it. The sensing coils allow just enough current to flow in the solenoid to maintain the float directly between the two coils.

If the density of the fluid increases, the float will start to rise. The change in position is detected by the search coils and results in a signal to an amplifier, causing an increase in the current to the solenoid. This increases the force of magnetic attraction on the float and restores it to its original centered position. This closed-loop system positions the float and the variation of the solenoid current is calibrated as a measure of fluid density.

When operating temperatures are expected to vary, temperature compensation using platinum resistance wire elements is used. This type of density sensor can be mounted on tank or pipeline nozzles, directly immersed in the process stream, or provided with a sample chamber for mounting in a bypass line. It should be

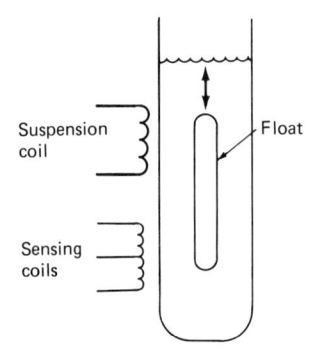

Figure 5-3. Electromagnetic suspension.

mounted within one degree of vertical, and it is important to avoid bringing magnetic materials closer than a foot (305 mm) from the float. The wetted parts can be epoxy, stainless steel, or gold plated for corrosion resistance.

Sample flow rates greater than a few gallons per hour or viscosities above a few centipoise can result in vertical forces on the float, causing measurement errors. For this reason this type of detector is not used in slurry service.

The available spans range from 0.01 to 0.4 specific gravity units within the limits of 0.4 and 2.0 SG. Measurement accuracies of 0.5 to 1% of full span are possible. The best results are obtained when the liquid density to be sensed is close to the density of the float. Pressure and temperature limitations are 200 psig (138 kPa) and 350°F (177°C).

FLUID DYNAMIC LIQUID DENSITY SENSOR

The fluid dynamic density meter can be used to measure the densities of gases and liquids. It has two chambers with supply nozzles to the fluid or gas. One chamber is used as a reference chamber with only a small outlet port. The dynamic pressure from this port serves as a reference value.

The measuring chamber has large inlet and outlet ports through which the measured liquid is pumped. The product to be measured affects the dynamic pressure of the jet on the receiver port inversely as a function of the density. A comparison of the pressure differential produced between the reference chamber and the measured pressure is a measure of the density of the product.

The fluid dynamic density meter can be used for the measurement of the density of process streams and effluents. It has no moving parts, a high sensitivity, and a high rate of response, but it is not particularly useful for non-Newtonian fluids. It can be manufactured from a wide variety of materials and is useful in petroleum and petrochemical refineries, natural gas processing plants, chemical process plants, pulp and paper processing, and in the manufacture of ethylene.

The torque tube type of displacer device used for level sensing can be used as a density sensor; however, several modifications are required. The sample fluid enters through a piezometer ring which eliminates the velocity effects of the flowing liquid. If the sample fluid velocity is below two feet (61 cm) per minute, the piezometer ring is not required.

The sample fluid exits through the top and bottom should be maintained at equal rates by the use of the purgemeters. If the density is to be measured in tanks or vessels, a flange-mounted displacer system can be used.

The displacer type of density sensor is available with a pneumatic output signal for intrinsic safety. Temperature compensation can also be built into the system.

HYDROMETERS FOR LIQUID DENSITY MEASUREMENT

Hydrometer devices use a weighted float with a small indicator stem at the top as illustrated in Figure 5-2. The stem is divided into density units. Archimedes' principle states that when a body is immersed in a fluid it loses weight equal to the weight of the liquid displaced. The hydrometer float is a constant-weight body, and if it is immersed in liquids with different densities, will displace a volume of liquid which is a function of the density.

The stem submersion is a measurement of the liquid density. The accuracy of the measurement depends on the surface tension, turbulence, and sample contamination.

An in-line density indicator of this type uses a chamber with an inlet and an outlet. The process sample enters the bottom and overflows to maintain a constant level. The sample flow rate is maintained at less than 1 gpm (0.0004 m^3/min) to minimize velocity and turbulence errors.

The hydrometer float can also be mounted inside a rotameter housing. In these designs, an overflow is also used to maintain a constant level of the sample. A needle valve is used to control the sample rate, which is usually about 15 gph (0.057 m^3/min). These units can operate at up to 200°F (93°C) and 100 psig (690 kPa). Specific gravity spans are 0.1 to 0.5 within the limits of 0.6 and 2.1 SG. Inaccuracy is about 1% of span.

The variable-immersion hydrometer units with a rotameter housing are also available in a servo-operated impedance bridge design. These units employ temperature compensation to correct the apparent fluid density for a predetermined base temperature.

Another hydrometer design uses capacitance in the form of a dielectric cup inside two insulated, concentric cylinders. The resulting change in capacitance is proportional to the density. Temperature compensation is provided so that the output signal is referred to 60°F (10°C). The specific gravity for these units must be between 0.7 and 0.9.

These hydrometer units are accurate, frictionless, without mechanical linkages to wear, and compatible with most corrosive fluids. Their major limitations are in the process liquids they can handle and the pressures and temperatures they can be exposed to.

The float position can be affected by velocity, friction, turbulence, and viscosity so these must be minimized. Material buildup on the float can also be a problem. These devices operate best in applications with clean nonviscous process fluids with a sample flow rate at the low gallons per hour level.

HYDROSTATIC HEAD DEVICES FOR LIQUID DENSITY

The hydrostatic head of a liquid column will vary as a function of the process fluid density. The approach to density measurement is the same as that used for

102 TRANSDUCERS FOR AUTOMATION

level detection with the following differences: (1) both sensing points may be located below the liquid surface; (2) the range of the sensor is suppressed.

This differential pressure approach to density measurement requires the same hardware as is needed for level measurement and the same precautions are needed in selecting equipment. The nature of the process fluid is important and the device may be affected by corrosion, plugging, contamination, and temperature effects.

Temperature compensation can be provided by the addition of temperature sensors as required by the measurement accuracy desired. The temperature effect of near-ambient temperature variations is relatively small for most liquids (Fig. 5-4).

The drawbacks and limitations of these detectors include the need for purge systems, which are recommended only for open tank applications. The speed of response will be slower because of the volume of the purge system, and in order to obtain accuracy, frequent maintenance will be needed. Other limitations include the maximum height of the liquid column and the resulting minimum span.

OSCILLATING-CORIOLIS LIQUID DENSITY SENSORS

The oscillating-Coriolis density sensor uses a C-shaped pipe and a T-shaped leaf spring which act as opposite legs of a tuning fork. An electromagnetic force excites the tuning fork, subjecting each moving particle within the pipe to a Coriolis acceleration. The resulting forces deflect the C-shaped pipe by an amount that is inversely proportional to the stiffness of the pipe and proportional to the

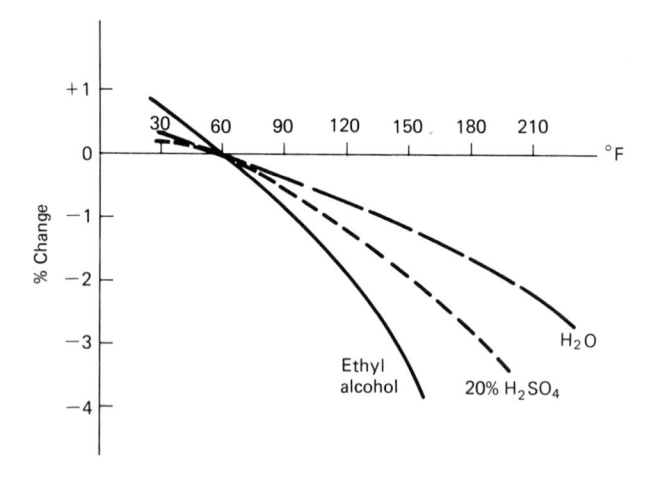

Figure 5-4. Temperature effects on density.

DENSITY AND WEIGHT MEASUREMENT

mass flow in the pipe. If the volumetric flow rate is held constant, then this deflection is a function of density.

The angular deflection of the pipe can be measured optically. This is usually done twice during each cycle of the tuning fork oscillation. The output can be proportionally modulated to the mass flow rate and an oscillator/counter circuit can be used to encode the pulse width data.

RADIATION LIQUID DENSITY SENSOR

Radiation liquid density sensors use a radioactive source beam and a detector system to measure the amount of transmitted radiation. As the gamma rays pass through the liquid, they are absorbed in proportion to the material density as shown in Figure 5-5. Any decrease in the density results in an increased output current since the liquid will absorb fewer gamma rays.

In pipe under 6 in. in diameter the radiation path is not adequate to provide high sensitivity. Therefore, an installation like that shown in Figure 5-6b is used to lengthen the radiation path and increase sensitivity.

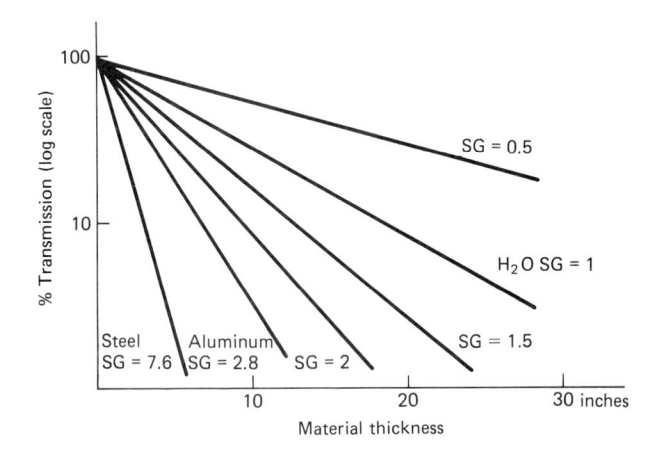

Figure 5-5. Cesium-137 transmission characteristics.

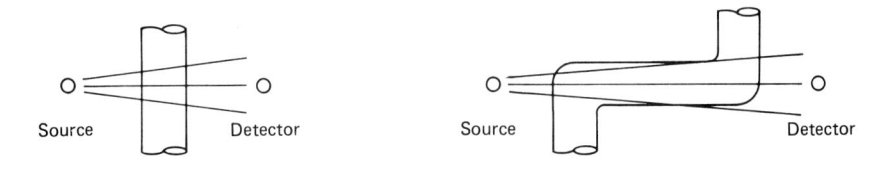

Figure 5-6. Radiation density sensing. *a*. Direct path. *b*. Lengthening the path.

The radiation source for most density applications is cesium 137. Source sizes normally range from 200 to 2000 millicuries as a function of the pipe diameter and specific gravity span. The gauge sensitivity can be increased by the use of collimated (narrow) beam geometry, which restricts the radiation in all directions except for a direct path to the detector. This minimizes scattering and allows the use of larger sources for increased measurement sensitivity.

Most units are designed so that the radiation intensity at a distance of one foot in any direction will not exceed five milliroentgens per hour. This is considered safe for any process area where the operator's occupancy is twenty hours or less per week. The rate at which the exposure is accumulated is also important. It is desirable to keep the yearly dose below 5 rems, and it should not exceed 12 rems per year or 3 rems per quarter. In most industrial applications it is possible to keep operator exposure below these levels.

For each industrial installation, it is important to estimate the dosage received by personnel working in the vicinity of the source, using an assumed occupancy figure and a "worst case" figure for proximity to the source.

Assume that the occupancy is 25 h per week and that the operator is within 12 in. (305 mm) of the tank during this period; the worst case is when the operator is working next to the source. Assume also that the holder meets the requirements for 5 mr/h; then the operator exposure is $5 \times 25 = 125$ mrem per week, or approximately 6.25 rem per 50-week year. This exposure would exceed set limits.

The second worst case would be if the operator were working by the detector when the tank was empty. Here the field intensity would be approximately 2 mr/h and operator's weekly and yearly exposure would be 50 mrem and 2.5 rem. This illustrates the point that the source shutter should always be closed when the tank is empty. Special interlock systems are available to prevent a maintenance worker from entering the tank when the source shutter is open.

In most cases, exposure can be minimized by positioning the detector and source so that the operator does not have to be in the area of high field intensity.

The source holder may use a shutter to close the radiation port. Some shutters employ a fail-safe design which automatically closes or a remote shutter control which allows closing the radiation beam from a control board.

The three basic types of gamma ray detectors are (1) the Geiger tube, (2) the scintillation detector, and (3) the ionization chamber.

The geiger tube measures radiation through the ionization of a halogen gas which is at about 500 V dc. It tends to have a low accuracy for density applications.

The scintillation detector senses the photons resulting from gamma ray interactions with certain crystal materials. This provides a sensitive detector which is greatly affected by variations in the environment, particularly temperature.

Ionization chambers operate by the ionization of pressurized gas between two dissimilar metals under incident radiation. The ionization generates a low current in the microamp range.

The ionization chambers are used in most density applications. They require stable amplifiers, but when this is provided, they are accurate and reliable. Since the output signal from the detector to the amplifier is in microamps the signal must be shielded to minimize noise pickup.

The detector chamber can be heated and controlled to eliminate temperature and moisture condensation problems. The process temperature is usually not critical, but it may be necessary to use thermal insulation to keep the detector temperature below 140°F (60°C) for some units. Absorbers inserted between the source and the detector with the pipe empty can be used to calibrate both zero and span. The minimum full-scale span is about 0.05 specific gravity units with a corresponding accuracy of 0.0005 or better.

For measuring small spans, the zero drift due to source decay becomes important and a source decay compensator may be needed. For installations with wider ranges, it may be required if a cobalt-60 source is used; the source decay with cesium 137 is less than 3% a year.

Beta ray absorption can be used for cryogenic density measurements. A strontium-90 source is used with a silicon surface barrier detector. The source and receiver are packaged in a small, corrosion-resistant probe and can also be used for level sensing.

Cost and safety considerations are the main limitations of the radiation density units. They are reliable once installed properly and offer precision for hard-to-handle process streams.

SOUND VELOCITY LIQUID DENSITY SENSOR

The sound velocity density detector uses the speed of sound in a liquid to measure the density. The speed of sound is given by

$$C = \sqrt{E/\rho}$$

where

E = bulk modulus,
ρ = mass density.

Since the velocity of sound is a function of both the bulk modulus and the density, any variations in bulk modulus will also cause measurement errors.

The liquid must be clear; emulsions, dispersions, and slurries scatter the sound waves and undissolved gases and bubbles cause errors. A common application of these sensors is in pipelines to detect the interface between two different hydrocarbon products.

This type of sensor can be calibrated by immersing the sensor in distilled water at temperature intervals between 32 and 167°F (0 and 75°C) and measuring the cycle repetition frequencies.

TORSIONAL VIBRATION LIQUID DENSITY SENSOR

The torsional vibration density sensor uses two cylinders mounted on a single shaft. The two cylinders are rotated in opposite directions to produce a torsional vibration. The density of the fluid changes the moment of inertia of the cylinders. The drive system is controlled by the signals from piezoelectric crystals. The crystals produce an output which is a function of the torsional strain of the shaft. This output is used to produce a resonant oscillating system. The frequency of resonance is a measure of the liquid density and can be calibrated in units of density or specific gravity.

VIBRATING-PLATE LIQUID DENSITY SENSOR

The vibrating-plate-type sensor uses a flexible rectangular plate fixed at both ends within a tube through which the liquid is flowing. A distributed mass of fluid of defined volume interacts with the plate, causing it to vibrate with a simple harmonic motion. The system is allowed to oscillate at its resonant frequency. A measurement of the period of oscillation is made and related to the density of the liquid.

This type of motion is harmonic, and the laws governing such motion define the velocity and acceleration of any point on the elastic member. The principle can be illustrated with the use of Figure 5-7:

1. A mass (M) with an established volume is suspended from a spring having a spring constant k (lb/in. deflection).

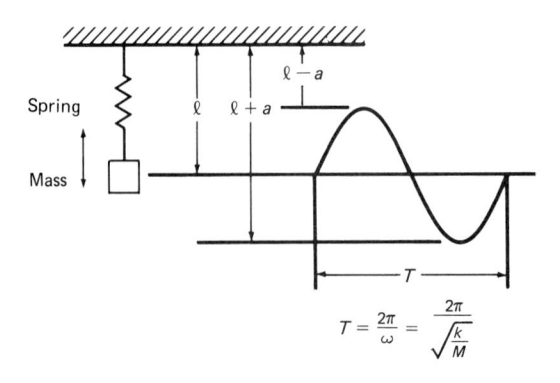

$$T = \frac{2\pi}{\omega} = \frac{2\pi}{\sqrt{\frac{k}{M}}}$$

Figure 5-7. Mechanical vibrations.

2. Assume that no motion exists and a spring length is established in which the tensile reaction of the spring equals the suspended mass.
3. An external force is applied reducing l to $l - a$.
4. The supporting force is now removed and the mass is permitted to fall to a position $(l + a)$.

In an undampened system, the mass will oscillate with a deflection of $\pm a$ distance with simple harmonic motion and a period

$$T = \frac{2\pi}{\sqrt{k/M}}$$

or

$$M = kT^2/(2\pi)^2,$$

where

T = period of oscillation,
k = spring constant,
M = suspended mass.

This implies that the measured density is directly related to the square of the period of vibration.

VIBRATING-SPOOL LIQUID DENSITY SENSOR

Vibrating-spool liquid density sensors use a cylindrical spool in the fluid. If a circumferential oscillation normal to the spool is induced and sustained, the spool will vibrate at a frequency that is a function of the spool's stiffness and the oscillating mass. Since the fluid surrounding the spool is caused to oscillate, the mass of the entire system in vibration consists of the mass of the spool plus that of the fluid. The system can be viewed as a lumped-parameter damped harmonic oscillator. For a closed-loop oscillator with a 90° phase shift, the vibration will be sustained at the natural frequency regardless of the fluid viscosity. Since the spool has a fixed stiffness and mass, variations in oscillation frequencies will be due mainly to variations in the fluid density.

In the system shown in Figure 5-8 the oscillations are induced and sustained by a feedback amplifier. A predetermined number of oscillations is counted and the elapsed time is measured by a clock.

Temperature effects due to changes in the spool dimensions and elastic modulus are minimized with the use of alloys like Ni-Span C. There is fluid pressure

108 TRANSDUCERS FOR AUTOMATION

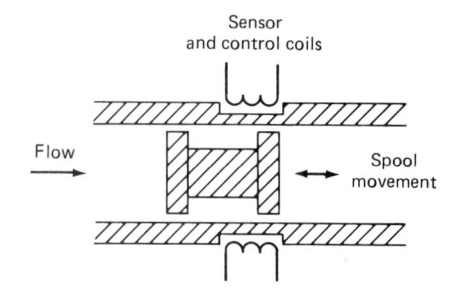

Figure 5-8. Vibrating-spool sensor.

on both surfaces of the spool, so pressure variations tend to have a cancelling effect. The oscillation amplitude is kept low so material stresses are minimal.

VIBRATING-TUBE LIQUID DENSITY SENSOR

These sensors excite a mass into mechanical vibration with a pulsating drive and then measure the amplitude of the vibration, which is proportional to the mass. This is illustrated in Figure 5-9. The process fluid flows continously through a U-tube section which is welded at the node points. The total mass of the U-tube assembly is affected by the process fluid density. A pulsating current through the drive coil forces the U-tube into mechanical vibrations. Any change in process fluid density causes a proportionate change in the effective mass of the U-tube and affects the corresponding vibration amplitude. An armature and coil pickup can be used to detect the vibration.

Automatic temperature compensation in the form of a resistance temperature element in the process stream can be used to perform temperature corrections.

The design pressures and temperatures are not limited by flexible connectors and the ambient temperature, process pressure, sample flow rate, or viscosity variations have little effect on the measurement.

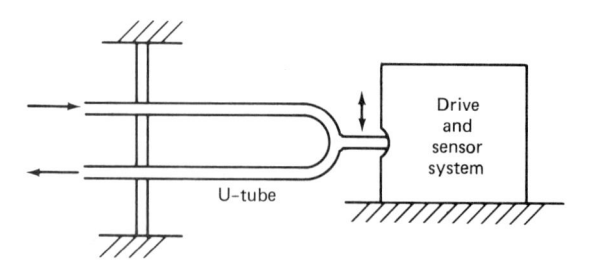

Figure 5-9. Vibrating-tube sensor.

DENSITY AND WEIGHT MEASUREMENT 109

This type of sensor can only handle clean fluids with low or moderate viscosities. High-viscosity streams or heavy slurries are likely to plug the U-tube.

When the process stream contains entrained gases, it is important to maintain the process pressure and flow, since low velocities can cause separation and trapping of the gas, and pressure variations can change the volume of the gas bubbles, affecting the density measurement.

One of the basic methods of weighing a fixed volume has already been discussed in reference to the displacement-type density sensors. Other constant-volume, variable-weight density sensors rely on this same principle. All of these use a constant-volume flow-through chamber which is continuously weighed. The design variations involve the shape of the weighing chamber, which can be in the form of a straight piece of pipe, a U-tube section, or a small tank.

WEIGHT-BULB DENSITY SENSOR

Weight-bulb density sensors use a bulb with sample tubes (Figure 5-10). The tubes also act as spring elements for the weighing mechanism. The bulb is suspended from one end of a beam and balanced by a counterweight on the other end. As the liquid density changes, the bulb will rise or fall, as allowed by the spring action of the sample tubes. The bulb motion is directly proportional to the density variations in the process fluid and is transmitted to the main beam, which also moves in proportion to the changes in specific gravity of the flowing sample. This type of sensor is also available as a motion balance controller using either pneumatic or electronic drives.

A thermostatic coil can be used to detect the outlet process temperature and correct for changes in specific gravity due to temperature variations. The compensator can correct for process temperatures between -100 and $500°F$ (-73 and $260°C$).

The main advantage of the weighed-bulb design is that it contains no flexible hoses so it can be used with high process pressures and temperatures. Since the

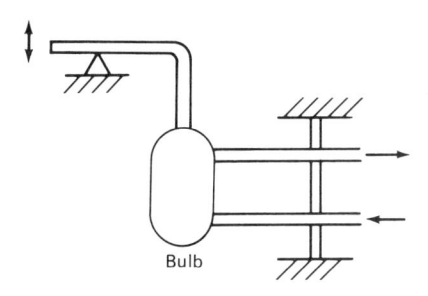

Figure 5-10. Weight-bulb sensor.

110 TRANSDUCERS FOR AUTOMATION

unit is designed for low sample flow rates with velocities in the laminar range, the forces created by fluid turbulence are low, but the small tubes and openings used are subject to plugging unless used for clean and nonviscous fluids.

U-TUBE DENSITY SENSOR

This sensor uses a U-loop pipe section pivoted about the horizontal axis (Figure 5-11). Process fluid flows through this loop and its weight is transferred to a weight beam. A counterweight balances the pipe section. A decrease in process density reduces the force on the weight beam, which is sensed by the nozzle of a force-balance controller. Because of the force-balance design, the total motion is a few thousandths of an inch. A dashpot may be used to reduce vibration effects or sudden changes in force.

This type of sensor can be used for slurries if the flow velocity is high enough to prevent settling. This usually requires velocities in the range of 5 to 8 f/s (1.5–2.4 m/s).

The process fluid can also contain gas bubbles. The process density without gas entrainment can be measured by installing a trap or separator upstream.

If the specific gravity of the total stream is to be measured, then the process pressure must be controlled so that the gas bubbles maintain their volume, and the fluid velocity must be high enough to prevent separation or trapping of the gas within the U-tube.

STRAIGHT-TUBE DENSITY TRANSMITTER

Straight-tube density transmitter devices are useful for applications where the process fluid is a heavy slurry or one that contains large solid particles which would not flow through a U-tube loop. The principle of operation is the same as for the U-tube design, but the performance is lower. The main advantage is

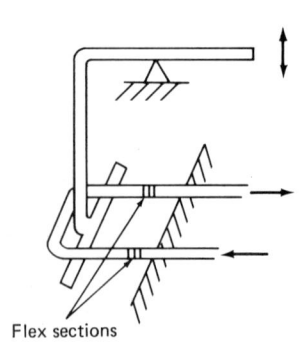

Flex sections

Figure 5-11. U-tube sensor.

that any process fluid that will flow through a 2-in. (50-mm) pipe will pass through this detector. The drawbacks include the fact that narrow spans are not available and, therefore, the ±1% inaccuracy based on span represents a substantially greater error than in the U-tube design.

DIRECT DENSITY CONTROLLER

The density sensors considered above use the counterweight for zero suppression only and the output is a function of calibrated spring movement or of the force generated by a rebalancing loop. The direct density controller uses the counterweight to balance the sample chamber and the liquid it contains at the desired density. The counterweight represents a controlled set point and any deviation in density will deflect the beam and initiate a counteraction.

GAS DENSITY SENSORS FOR OPERATIONAL CONDITIONS

These sensors detect the process gas density at operating conditions and generate a signal which can be used for mass flow rate calculations with the orifice or volumetric flow data. The two basic designs for these detectors are based on displacement or centrifugal techniques.

DISPLACEMENT GAS DENSITY SENSOR

Displacement gas density sensors measure the gas density at actual flowing conditions using Archimedes' principle, which states that the buoyant force on a float is a function only of the fluid density surrounding the float. The effects of process pressure, temperature, supercompressibility, and specific gravity changes are automatically detected in these devices.

One version of these sensors uses a float connected to one end of a pivoted beam with a temperature-insensitive spring connected to the other end (Figure 5-12). The spring tension is adjusted until the beam reaches the null position. A measurement of the spring displacement provides linear density data.

The spring tension can also be adjusted by pneumatic relays or electric motors.

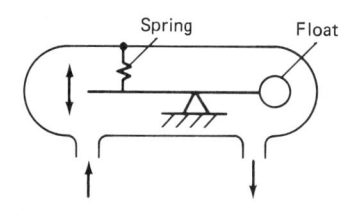

Figure 5-12. Displacement gas density sensor.

In these cases, when a process density change takes place, a proportional change occurs in the force acting on the float which causes the controller to adjust the tension of the spring until the connecting rod is balanced again. The equilibrium position is proportional to the process density.

In the pneumatic devices the process gas pressure must be above 5 psig (34 kPa) and the gas must be clean or it might plug the pneumatic system.

For accurate measurement the same operating conditions must be present in the detector chamber as in the process line. This is achieved by the use of short (close-coupled) sample lines and high sample flow rates (10 scfh or 280 L/h). Spans from 1.0 lbm/ft to 25.0 lbm/ft (16 to 400 kg/m) can be obtained for measurements at 0.25% accuracy.

CENTRIFUGAL GAS DENSITY SENSOR

The centrifugal sensor measures the gas density at flowing pressure and temperature conditions using a small centrifugal blower which extracts a small gas sample from a tank or pipe line. The impeller operates at a constant high speed of 3000 to 13,000 rpm. Gas enters the impeller at the center and is thrown outward by centrifugal force. This creates a pressure differential across the impeller which is proportional to the gas density. The differential pressure is then measured with a pressure transducer.

Errors can be introduced by temperature differences between the process and the density chamber. This temperature difference may be caused by the difference between ambient and process temperatures and also by the motor and friction heat developed in the sensor. The error can be as high as 0.1 or 0.2% per degree Fahrenheit. In most installations, close-coupling and insulating the detector are sufficient. In case of critical measurements, the process and chamber temperatures are detected and automatic correction is applied for the difference.

If the gas sample is dirty, deposits can accumulate in the unit, affecting the sample flow rate.

APPLICATION AND LIMITATIONS

The compressibility of most liquids is small and the effects of pressure upon density measurements can usually be disregarded. But where operating temperatures vary, it is necessary to compensate for these changes. All density detectors can be furnished with some means of temperature compensation.

Most density units are limited in use to clean, nonviscous fluids. If the process fluid is viscous or of the slurry type, then the radiation, the hydrostatic head or the U- or straight-tube sensors must be used. Of these three types, the U- or straight-tube gauges are limited in their pressure and temperature ratings and can be used for only moderately viscous or slurry-type streams.

The hydrostatic head type of sensor makes it difficult to detect narrow spans because the corresponding height of the required pipe becomes excessive. In heavy slurry or viscous services, the operating pressure is limited by the rating of the pressure repeater or by the pressure at which purge media are available. The existence of slurry or high-viscosity processes with high operating pressures requires the use of radiation-type sensors.

In the selection of gas specific gravity sensors (composition detectors), costs favor the displacement type of indicator. For continuous indication or remote signal transmission, the gas column balance and the viscous drag-type instruments are the most economic choice. If the major considerations are high sensitivity and accuracy, the electromagnetic suspension-type sensor appears to be the best choice. If corrosive gas samples are involved and the sensing elements should not come in contact with the process stream, the thermal gauges may be best. These units all operate at near-ambient conditions and if nonambient samples are used it is necessary to regulate the pressures and temperatures as well as the moisture contents of both the sample and the reference gas, which is usually air.

If the purpose of measurement is direct density detection under operating conditions, the best choice may be the manually operated displacement sensor. Continuous indication or remote readout can be provided by both the displacement and the centrifugal-type designs. These units are accurate and capable of operation at high pressures and temperatures.

An important consideration is to provide the gas sample without pressure or temperature losses. This can be achieved by close-coupling the components to the process and insulating the sample lines. Devices with restrictions and moving parts should be protected from deposits by filtering the sample.

WEIGHT SENSORS: APPLICATION AND SELECTION CONSIDERATIONS

In some manufacturing operations and process control systems, gravimetric (weight) control can provide the most direct and accurate measurement. Suppose, for example, we wish to maintain constant proportions between solvent and solute in a process that the solvent may enter at various temperatures. If volumetric measurements of the liquid solvent is used, constant temperature correction must be applied, but if the ingredients are controlled by weight, then a more direct control is maintained and more accurate measurements are possible.

Many weighing applications involve the batch mixing of various materials and the use of weight scales can require the purchase of special tanks and pumps. A comparison can be made with flow metering hardware that requires no additional equipment. The weighing method is preferred in cases where the me-

114 TRANSDUCERS FOR AUTOMATION

dium is hard to handle, as in solids, slurries, and high-viscosity or very hot fluids. In most other cases, the choice can be made on the basis of installed cost, accuracy, and other important considerations.

The weight sensor must be selected considering the features and the nature of the plant involved. In the case of a system of centralized controls with computer supervision, the best selection may be the strain gauge. Generally, one should consider the measurement technique that requires the least number of signal conversions.

In the handling of many liquids and solids, weighing techniques provide a more accurate and direct method of control which is relatively independent of process medium characteristics, physical location, time, atmospheric conditions, and temperature. Thus, units of weight are easily converted to monetary values, facilitating their use in commerce and exchange.

In the handling of dangerous or highly corrosive materials the weighing process provides a safe and convenient method of control without container penetration or other problems. It can also be used for cryogenic materials.

Compared to volumetric measurement, weighing has several advantages. The most basic of these is freedom from temperature effects. In volumetric measurements, the temperature of the material can cause volume changes and require correction factors to be applied. Weight is not affected by temperature, and if the material can be physically handled, it can be weighed.

In the weighing of molten metal, temperatures of 2400–2600°F (1300–1400°C) are common; however, accurate weighing to 0.1% is possible with strain gauge weighing equipment.

SPRING BALANCE SCALES

The spring balance scale uses a calibrated spring and a means of reading the deflection or elongation of the spring under load. This technique can have a high accuracy when properly designed with high-grade constant-temperature modulus alloys.

Spring balance scales depend on a relatively large deflection of the weighed load and are best suited for relatively light loads. However, they do allow a low-cost approach to the measurement of weight.

MECHANICAL LEVER SCALES

Mechanical lever scales utilize levers which balance the weight of the unknown against a known mass, balancing one moment against another. The level system is usually adjusted so that the pull from the unknown is within a convenient range. The unknown mass can include the mass of the bin, hopper, or platform holding the material to be weighed. This is balanced out by a tare device.

The most common balance devices are the beam and the pendulum. The beam is made up of several smooth bars. One beam is for balancing out the tare weight and it may not be calibrated. Another may be used for balancing out hundreds or thousands of pounds and another may be used for tens and units. The total moment exerted by the beam is the sum of these. The balance is indicated by the position of the free end of the beam.

The pendulum is a heavy mass on a horizontal pivot. The moment is proportional to the displacement of the mass. Tare corrections are made with a tare beam.

A large number of weighing installations are based on mechanical lever scales. These include such applications as motor vehicle scales, railroad car scales, hopper scales, tank scales, platform scales, crane scales, dynamometer scales, and other specialized types.

These all utilize the principles of the lever and involve various combinations of load support systems. Portable models are used for the lower capacities, while the larger units are more permanent and usually require preparation of the installation area.

Mechanical lever scales have evolved through many years of development to their present accuracy and reliability. The performance of mechanical lever scales has been the basis for all laws and standards established in connection with weights and measures for the regulation of commerce on local, national, and international levels. Their basic simplicity allows them to be used even in those parts of the world where sophisticated maintenance service is not available. (See Table 5.2)

Mechanical lever scales may have almost any capacity, and can accommodate loads of almost any material. Overhead track scales, motor truck scales, and railroad track scales are all forms of mechanical lever scales. Vehicles in motion such as trucks or railroad cars can be weighed on mechanical lever scales if their velocity is less than 5 mph (8km/h). These scales usually employ pendulum counterbalances.

TABLE 5.2 Typical industrial scales.

Type of scale	Capacity
	(lbm/kg)
Even-arm	5/2.26
Bench dial	200/90
Platform	1500/680
Floor	6000/2720
Overhead	12,000/5443
Suspended hopper	25,000/22,345
Truck	100,000/45,400
Railroad track	400,000/181,800

Many processes require the weighing of batches of material individually while others may require the weighing of a series of materials for later mixing or other treatment. Batches of constant size can be weighed in a hopper or bin, with the correct weight indicating when the feed should be stopped. A series of quantities of materials can be weighed in this manner using photoelectric or other pickups to sense the balance beam position.

Granular materials can be weighed on conveyor-belt scales. A section of conveyor belt is built into the scale mechanism, with a tare adjustment to balance the scale when the belt is empty. Conveyor-belt scales are often used for supplying granular material at a known rate. The feed rate to the belt is controlled by a balance device and this keeps the load on the belt constant. The rate of feed of the material is then controlled by adjusting the speed of the belt.

Another application is the integrating weighing device. In this system, the belt is driven at a known speed and the total amount of material delivered is computed from the duration of the operation and the average weight of material on the belt. The amount of material on the belt should be uniform and constant.

Mechanical lever scales can provide long-term accuracy with the proper maintenance. They are also resistant to most of the usual environmental conditions. They are available in a wide range of capacities and forms. Sizes range from very small units to railroad track scales over 100 ft (30 m) long. They can easily be incorporated as a working part of other industrial devices.

The major limitation is in the speed of response. The mass and inertia of the lever system provide slower weighing speeds compared to strain gauge load cells, which may be used to weigh vehicles moving at high speeds.

Many electrical output devices can be used. The simplest of these are switching devices, which indicate when a desired weight or series of weights has been reached. Transducers can be used to provide a continuous electrical output which can be fed into a microcomputer control system.

HYDRAULIC LOAD CELLS

Hydraulic load cells operate on the principle of a force counterbalance. The weight on the load cell causes a change in internal fluid pressure. A number of pressure sensing devices are used to translate this pressure into a signal proportional to the weight.

The supported load is usually borne by a top member, or head plate, which rests upon a ball or rolling member. The rolling member is supported by a load-sensitive piston or column.

The basic design uses a close-fitting piston and cylinder with an O-ring to prevent leakage around the piston. While this can be satisfactory for some applications, there are some frictional losses due to the rubbing of the O-ring.

The rolling-diaphragm hydraulic load cell is shown in Figure 5-13. Here the hydraulic fluid is confined within the diaphragm chamber using a clamped seal

DENSITY AND WEIGHT MEASUREMENT   117

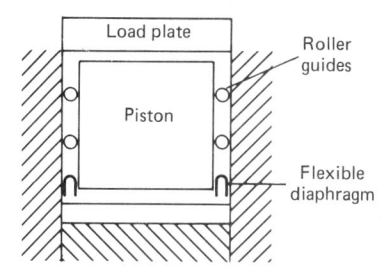

Figure 5-13. Rolling-diaphragm hydraulic load cell.

between the cylinder wall and base plate. The piston or load-bearing member is guided within the cylinder by guides. Piston travel is limited and the design provides the linearity and precise performance desired. One limitation is the ability of the diaphragm to withstand pressure. The materials available limit the maximum internal pressure to about 1000 psig (5.5 to 6.9 MPa).

The rolling-diaphragm type of hydraulic load cell can be used for most process weighing applications. Measurement inaccuracies of ±0.25% FS are possible.

An application advantage of the rolling-diaphragm load cell is its relative insensitivity to the amount of hydraulic filling. Thus, the connections can be via high-pressure hoses rather than rigid tubing. Changes in hydraulic fluid volume due to ambient temperature variations will also have little effect on accuracy.

In automation applications, hydraulic load cells must have the following characteristics: (1) they must function without leakage; (2) they must be relatively free of internal friction; (3) they must be linear and precise in operation.

The all-metal designs are characterized by an all-metal construction and use only a small amount of hydraulic fluid. The type of design allows extremely heavy unit loads. Load cells of this type can be used for capacities of up to 10,000,000 lbm (4,500,000 kg).

By eliminating bearings, pivots, and knife edges, hydraulic load cells can produce a high sustained accuracy. The displacement under load can usually be limited to 0.005 to 0.010 in. (0.125 to 0.25 mm).

Hydraulic load cells are self-contained and require no outside power. They are inherently explosion proof and units are available for both tension and compression applications. Most of the cells are applicable to tank, bin, and hopper weighing.

Hydraulic load cells are useful for high-impact loading and can withstand overloads of 300–400% without loss of accuracy or null point. The natural frequency of these cells is very high and resonance due to dynamic loads is rare. The viscous damping due to hydraulic action tends to produce stable signals even under dynamic conditions.

118 TRANSDUCERS FOR AUTOMATION

The high natural frequency, low deflection, and fast response rate of hydraulic load cells allows them to be adapted to such applications as web tension control, dynamometer torque measurement, and jet engine and rocket thrust measurement. Load cells and cell weighing systems have several advantages: (1) they are not subject to wear; (2) they are easy to install; (3) they have a fast response and a short settling time; (4) they provide a weight signal that is compatible with automation system requirements.

Temperature compensation is available for both span and zero errors. Standard operating limits without temperature compensation are 0 to 120°F (−18 to 49°C).

HYDRAULIC TOTALIZERS

Many hydraulic load cell applications involve process weighing and the containing vessel is normally supported with more than one load cell. The total weight is the output of all the load cells. If the load cells are just interconnected and an average pressure is obtained, there is the danger of grounding one cell, especially under conditions of nonuniform loading. This problem is eliminated through the use of a hydraulic totalizer. Here the output of each load cell is conducted to individual modules, which are made up of small pistons and cylinders. The output forces of these piston/cylinder combinations are transferred to an output module. This can usually be accomplished without serious internal losses and one pressure signal proportional to the inputs is developed.

Hydraulic totalizing inaccuracy on the order of 0.5% is common due to temperature sensitivity and other nonlinear effects, but units with totalizing inaccuracy of ±0.1% FS can be obtained.

Hydraulic load cells can also be totalized electronically by converting the hydraulic pressure outputs into a proportional signal. This method has the advantage of transmission without loss of accuracy.

PNEUMATIC LOAD CELLS

Pneumatic load cells can also be applied to process media weighing. Most units employ force-balance principles which allow high performance. The pneumatic pressure signal can be transduced into analog or digital form or transmitted by metal or plastic tubing to a remote point.

Pneumatic systems have a slow rate of response to incremental load changes but the force-balance technique keeps the deflection under load also low—0.003 to 0.005 in. (0.075 to 0.125 mm).

Since the natural frequency of pneumatic load cells is low, under some conditions of dynamic loading resonance can occur. This has been overcome in some designs with the use of stabilizing or damping chambers.

The major advantages of pneumatic weighing systems are (1) they are relatively insensitive to temperature variations; (2) they are explosion-proof; (3) they contain no contaminating fluids, which in the event of rupture or leakage could prove to be a problem in food and drug processing.

Pneumatic load cells must have a carefully regulated source of clean, dry air. Systems can be operated for short periods on inert gases such as dry nitrogen but this is impractical for most automation applications. An average requirement for the system air supply is 10 scfm (0.0047 m/s) of dry air [$-40°F$ ($-40°C$) dew point] per load cell.

STRAIN GAUGE LOAD CELLS

Strain gauge load cells are common in the weighing of tanks, bins, hoppers, ladles, platforms, rails, and conveyors. In a wire strain gauge, a wire is bonded to a supporting member such that its cross section varies as the supporting member. As the wire is strained, it is possible to establish a relationship between the electrical resistance of the wire and the force causing the deformation of the supporting member. In a strain gauge load cell the supporting member usually consists of one or more steel columns mounted on a rigid base and hermetically sealed within a case as shown in Figure 5-14.

Some of the first uses of the bonded resistance wire strain gauge were in load cells and force transducers. It is now practical to consider the use of strain gauge load cells in installations requiring inaccuracies of $\pm 0.1\%$ to $\pm 0.25\%$ of full scale. There are a wide variety of strain gauge of load cells available for many applications.

Strain gauge load cells can be used for compression or tension load measurements and some can be used in either direction. Load cells with two sets of strain gauge elements in a double-bridge configuration are also available.

The strain gauge load cell has been successfully applied to such weighing processes as automatic batching, automatic inventory recording and control,

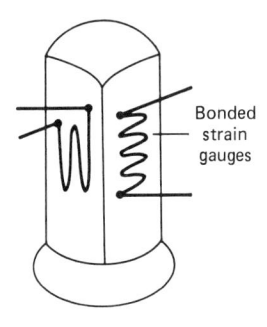

Figure 5-14. Strain gauge load cell.

constant-rate feeders, and continuous belt weighers. When strain gauge load cells are installed under tanks, bins, or hoppers which are subject to excessive bending expansion or contraction, special mounts may be required to isolate the load cell from external side forces.

In some applications there may be loading conditions or forces which create angular loads on the cell. Strain gauge load cells should be protected from these angular or nonaxial loads. These forces can result in a bending moment on the support column which is sensed by the strain gauge as a normal load. The support column can be stabilized to minimize this effect.

The output signal of a strain gauge is relatively small and is related to the excitation. A typical value is 2-3 mV per volt of excitation. The excitation voltage (ac or dc) is usually in the range of 5 to 20 V.

Strain gauge load cells must be compensated for the effects of temperature on zero shift and span. The strain wires are usually made of temperature-insensitive alloys and compensating resistors are used in a bridge configuration. A typical strain gauge circuit is shown in Figure 5-15.

Installations outside of normal temperature limits, which are usually 15 to 115°F (−9 to 46°C), require the use of correction factors or a means of controlling the temperatures around the load cells.

Strain gauge load cells must be used within their capacity ratings. Overloads in excess of ratings can result in loss of accuracy of device failure. Typically, an overload should be less than 125% of the rated capacity. This includes impact or shock loading, as well as static loading.

The successful application of strain gauge load cells requires attention to the

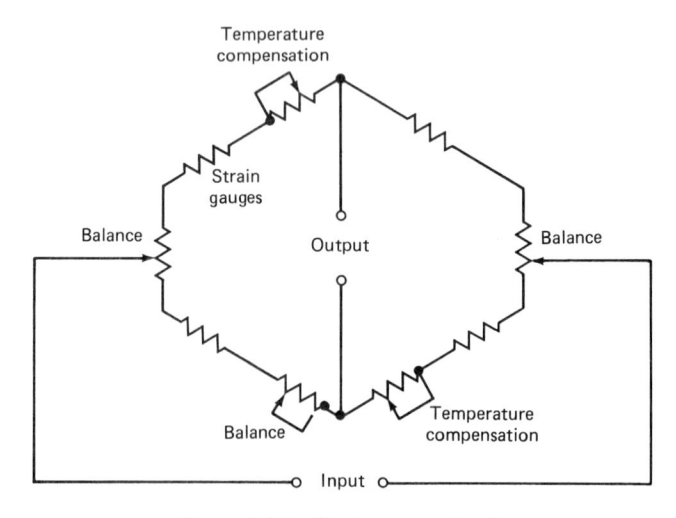

Figure 5-15. Strain gauge circuit.

complete system including the excitation power supply, output amplifier, cabling, and other equipment, such as filters.

SEMICONDUCTOR STRAIN GAUGES

The piezoresistive characteristics of germanium and silicon semiconductor materials are highly sensitive to applied stress or strain. Their gauge factors, the unit change in resistance divided by the unit strain, are more than fifty times those of metallic wire or foil strain gauges. In addition to a very high strain sensitivity compared to metallic strain gauges, they also possess relatively high nonlinearity, temperature effects, and terminal resistance. Semiconductor strain gauges are used in force measuring services where a high output and low system cost are desired.

SEMICONDUCTOR STRAIN GAUGE LOAD CELLS

Semiconductor strain gauge load cells are mostly used in systems requiring low cost and moderate accuracy. Semiconductor strain gauges in load cell configurations provide units with rated output capabilities of 1.0 V at 15 V bridge excitation. As a result of the high signal level, semiconductor units are used in simple weighing systems with simple regulated power supplies and direct outputs. A comparison of semiconductor and metallic strain gauge load cells is given in Table 5.3.

NUCLEAR RADIATION SENSORS

The nuclear radiation sensors are generally used for the weighing of bulk materials in motion. A radioactive source of gamma rays is directed at the moving

Table 5.3. Comparison of semiconductor and metallic strain gauge load cells.

Type of load cell	Semiconductor	Metallic
Output for 15-V input	1.0 V	30 mV
Linearity	0.25%	0.05%
Hysteresis	0.02%	0.02%
Temperature effects		
Zero	0.003%/°F	0.002%/°F
Output	0.006%/°F	0.001%/°F
Cost factor	1.2	1.0

material. The material absorbs some of the gamma rays and allows others to pass through. The amount of radiation transmitted through the bulk material depends on the amount of material on the conveyor. A radiation sensor detects the transmitted radiation and produces a signal which is proportional to the amount of material on the conveyor.

The nuclear radiation form of weight sensing is useful when the weight sensor should not contact the material or the conveying devices. The same safety precautions as discussed earlier in this chapter for radiation density sensors and in Chapter 7 for radiation level sensors should be employed. A number of problems of conventional belt scales can be avoided with this method.

INDUCTIVE AND RELUCTANCE LOAD CELLS

Inductive and reluctance load cells both use the motion of a ferromagnetic core (inductive) or a coil assembly (reluctance) which is converted to a voltage directly proportional to the displacement. Various force sensing elements are used to convert the applied force to a displacement to which the sensing element is coupled. These transducers provide a relatively high output signal with moderate to high accuracies for a broad range of measuring capacities.

INDUCTIVE SENSING TECHNIQUES

Inductive weight sensors use the change in inductance of a coil with the changing position of an iron core. The basic forms of inductive sensing are shown in Figure 5-16. In *a*, a motion of the iron core increases the inductance of coil 1 and decreases the inductance of coil 2. The two coils can be arranged as one half of a bridge network to develop a voltage proportional to the core position.

Inductive weight sensors supply a relatively high output signal and exhibit good stability. Their inertial mass is greater than that of strain gauge sensors, which makes them more sensitive to vibration.

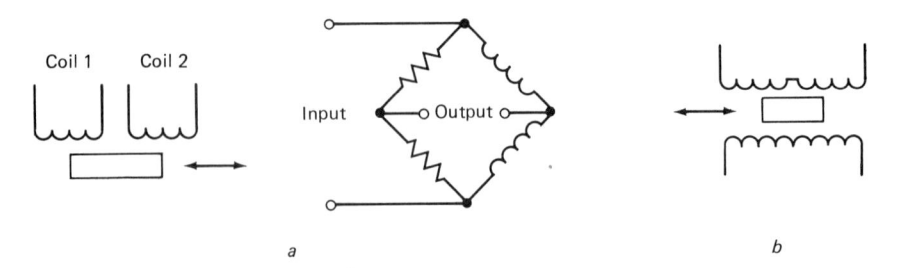

Figure 5-16. Inductive sensing. *a.* Two-coil system. *b.* Three-coil system.

VARIABLE-RELUCTANCE TECHNIQUES

This weight sensing technique is similar to inductive sensing except that the inductance of one or more coils is changed by altering the reluctance of a small air gap, as shown in Figure 5-17. The coils are mounted on a fixture of ferromagnetic material and an armature completes the magnetic circuit through the air gaps. A movement of the coil assembly to the right decreases one air gap while another is increased. The two coils can be connected in a bridge configuration for sensing the voltage proportional to the movement of the coil assembly.

The reluctive sensing technique provides a relatively high output voltage and good stability along with a high vibration sensitivity due to the high inertial masses of the structure.

Inductive and reluctive load cells generally exhibit the following characteristics:

Capacity—0.01 to 100,000 lb (0.0045 to 45,000 kg)
Output—5 to 200 mV/V
Linearity—0.1%
Repeatability—0.05%
Temperature effects
   Zero—0.001 %/°F
   Output—0.001%/°F

MAGNETOSTRICTIVE SENSING TECHNIQUES

These methods are based on the change in permeability of ferromagnetic materials with applied stress. A stack of laminations is used to form a load-bearing column and primary and secondary transformer windings are wound on the column as shown in Figure 5-18.

In the unstressed state, the permeability of the material is uniform throughout the structure. The coils are oriented at 90° with respect to each other, so very little coupling exists between them and the output signal is a minimum. As the

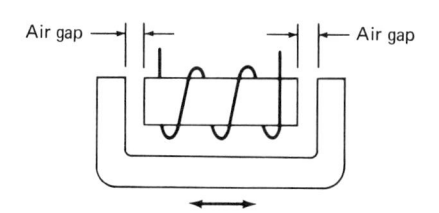

Figure 5-17. Variable-reluctance sensor.

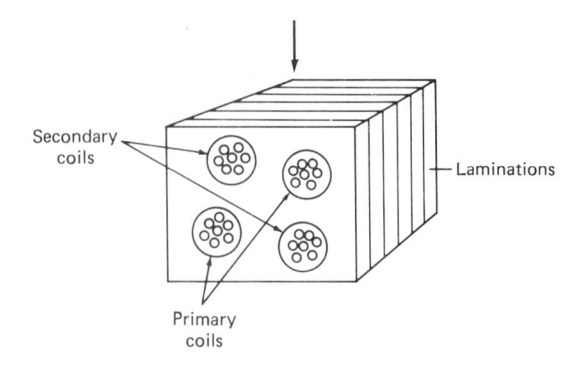

Figure 5-18. Magnetostrictive sensor.

column is loaded, the stresses created cause the permeability of the column to be nonuniform, producing distortions in the flux pattern of the magnetic material. A magnetic coupling now exists between the two coils and a voltage is induced in the signal coil, providing an output signal proportional to the applied load.

The magnetostrictive technique provides a relatively high output signal in a rugged load cell design. Several configurations can be used. One configuration is for applications in which there are no bearing surfaces on the devices being weighed; in the presence of lateral loads the unit is very sensitive and should be protected.

BEAM-TYPE STRAIN GAUGE TRANSDUCERS

Another type of force measuring transducer used for weighing depends on a slotted bending beam construction (Fig. 5-19). The strain gauges are arranged in a bridge configuration so that the output of the sensing element is independent of the position of the applied load. Electrical compensation is used to reduce the load position sensitivity.

The strain gauge location and beam design provide inherent adverse loading sensitivity. The simplified load sensing configuration provides inherent linearity

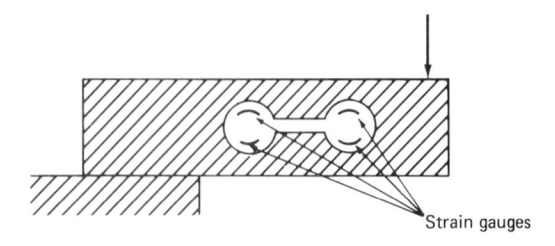

Figure 5-19. Beam strain gauge sensor.

DENSITY AND WEIGHT MEASUREMENT

as well as low creep. Beam force transducers generally exhibit the following characteristics:

Capacity—10 to 5000 lb (4.5 to 2270 kg)
Linearity—0.03%
Hysteresis—0.02%
Repeatability—0.02%
Creep—0.04%/h
Temperature effects
   Zero—0.002%/°F
   Output—0.001%/°F

MONORAIL WEIGHING TRANSDUCERS

Most of these devices are beam-type weighing transducers such as those used in the monorail conveyor systems in meat processing plants. Coventional monorail weighing systems support the live rail by load cell and flexure assemblies. These monorail transducers simulate the live rail and are self-supporting.

A typical device is shown in Figure 5-20. Strain gauges sense the bending strains as the load moves along the transducer. The gauges are connected in a bridge configuration for a constant output which is independent of the load position on the transducer.

This type of transducer allows a greater measuring accuracy and eliminates the second load cell needed in most conventional arrangements, reducing the overall system cost.

THE DIRECT WEIGHING OF TANK LEGS

Some structures may be already fabricated and erected, and their support by load cells can require extensive modifications. One technique used in weighing these structures is to install the strain gauges directly on the supporting legs (Fig. 5-21). The legs then become the transducer sensing elements which are

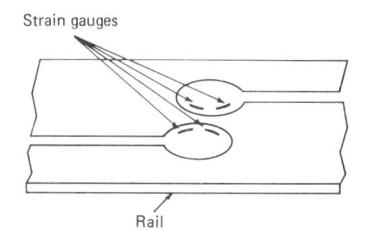

Figure 5-20. Monorail weight sensor.

126 TRANSDUCERS FOR AUTOMATION

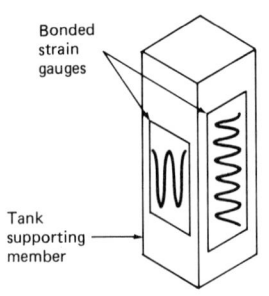

Figure 5-21. Using a strain gauge load cell as a supporting tank leg.

connected in a bridge configuration. One pair of gauges is applied longitudinally, sensing the compressive stresses in the tank legs. Another pair is applied in the transverse direction, sensing the tensile strains. The four gauges are connected in a bridge arrangement which is protected with waterproofing materials. The strains in the supporting legs are usually low and it is difficult to maintain a perfect waterproofing. This results in an accuracy of 3 to 5% for these systems.

HIGH-TEMPERATURE LOAD CELLS

In the metal processing industry, the need for devices to withstand high environmental temperatures is great. Organic and inorganic bonded strain gauge backing and installation materials which are now available can withstand higher temperatures than most conventional units. Bonded strain gauges with organic backings can operate continuously at temperatures of up to 500°F (260°C).

High-temperature strain sensing wire alloys can also be used with organic bonding materials. Flame spraying techniques can be used where molten aluminum oxide is sprayed on the sensing element. These techniques permit short-term operation at temperatures of up to 1000°F (538°C) but with some decrease in performance.

EXERCISES

1. Define the density of a substance and its relation to specific gravity.
2. Discuss how density may be obtained indirectly.
3. What is supercompressibility?
4. What are the limitations of angular position density sensors?
5. Discuss the operation of the ball-type density meter with the aid of a diagram. What is the major advantage of this type of sensor?
6. Where is the capacitance type of liquid density sensor most useful?
7. What conditions must be avoided when using displacement float density sensors?

DENSITY AND WEIGHT MEASUREMENT

8. Discuss the operation of the electromagnetic suspension density sensor. What are the installation requirements?
9. What are the advantages of the fluid dynamic type of density sensor?
10. Discuss how a hydrometer device could be provided with an electrical output.
11. What are the safety requirements in using radiation density systems?
12. Discuss the principles of operation of sound velocity density meters. What are the medium requirements?
13. What vibration principles can be employed to measure density?
14. Discuss how displacement and centrifugal gas density sensors operate. What are the requirements for accurate measurements?
15. What are the advantages of weighing over volumetric measurements?
16. Discuss the automatic weighing of material on a conveyor belt.
17. What applications are suitable for hydraulic load cells?
18. Where are hydraulic totalizers used?
19. Discuss the advantages of pneumatic load cells.
20. What considerations are important to the application of strain gauge load cells?
21. Discuss the differences between metallic and semiconductor strain gauge load cells.
22. How do magnetostrictive load cells operate?
23. What are the differences between beam and monorail force sensors?
24. What techniques can be used for high-temperature weighing applications?

6
VISCOSITY MEASUREMENT

Viscosity should be measured at flow rates low enough for the fluid to move in layers, for if the flow is turbulent, the measurement must be corrected. The unit of viscosity is the poise (P). This is defined as the force required to move one of two parallel planes, each one square centimeter in area, a distance of one centimeter at a velocity of one centimeter per second, when the planes are separated by one centimeter of fluid. Molasses, water, and air have the approximate values of 2000, 0.01, and 0.00018 P. Most liquids have low viscosities and the centipoise (0.01 P) is a more convenient unit. Another advantage is that the viscosity of water at 68.4°F is 1 cP.

A basic technique of measuring viscosity involves dropping a ball bearing through a column of viscous material in a tube and measuring the time it takes for the ball to reach the bottom. Another method used to measure the viscosity of a fluid is by sending air bubbles up a sample column and timing the rise of bubbles through the fluid.

These techniques are methods of viscosity measurement that involve Stoke's law and are not valid except for flows at very low Reynolds numbers, which are found in viscous oils or gases of high hydrogen content. The error may be large unless frequent calibration runs are made.

Viscosity can be measured on an intermittent basis using samples or continuously in a system. The intermittent measurements usually are made by laboratory instruments.

If viscosity is to be controlled under dynamic conditions in an automation system, then a continuous-flow viscometer is needed. Transducer selection is based on the range of viscosity and whether or not the medium is a Newtonian fluid.

A Newtonian substance, if it is subjected to shear stress, undergoes a deformation such that the ratio of the shear rate or flow to the shear stress of force exerted is a constant. A non-Newtonian substance does not have a constant ratio of flow or force.

PRESSURE DROP MEASUREMENT TECHNIQUES

A viscosity measuring system for continuous automatic control may use pressure drop techniques. The liquid being measured is pumped through a restricting tube or orifice plate at a constant rate and temperature. The viscosity can then be measured using a viscosity-sensitive rotameter which is calibrated in centipoise units.

In the use of an orifice plate to obtain a pressure drop to measure viscosity, the Reynolds number plays an important role. The Reynolds number provides an index of the ratio of inertial to viscous forces. Inertial forces dominate when the Reynolds number is high, and the coefficient of the Venturi tube, the flow nozzle, or the orifice plate becomes a constant for most practical purposes.

If the viscosity may change, the resulting change in the coefficient of the restriction device affects the flow measurements for low Reynolds numbers. The volume V passing through the pipe per unit time is proportional to the difference in pressure $(P_1 - P_2)$, the fourth power of the radius (r), and to the reciprocal of the length of the tube (l) and the coefficient of viscosity (n) as shown below:

$$V = \frac{r^4(P_1 - P_2)}{8nl}.$$

The continuous capillary viscometer makes use of this relation to convert the differential pressure into viscosity measurements.

A continuous capillary viscometer is shown in Figure 6-1. It uses a sample flow rate of about 1 gph. The measurement span is a function of the bore and length of the capillary, which allows measurement of a variety of viscosity ranges. Most continuous capillary viscometers are used to measure the viscosity of Newtonian liquids. The pressure transducers that are used to transmit the measured viscosity allow it to be adaptable to automatic computer control.

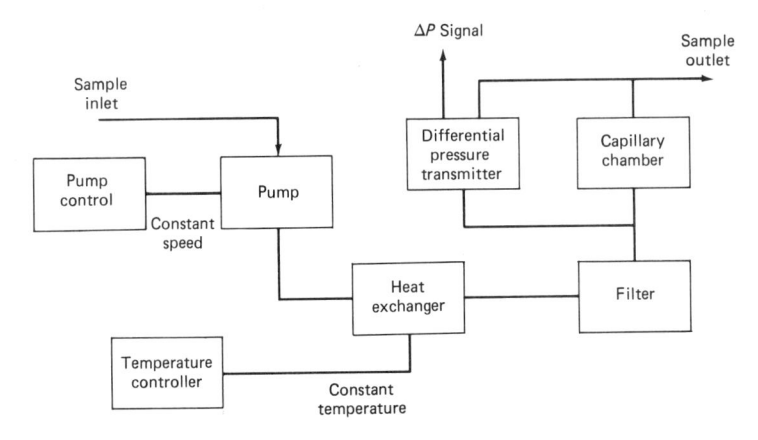

Figure 6-1. Continuous capillary viscometer.

130 TRANSDUCERS FOR AUTOMATION

This type of viscometer can be used for viscosities to 5000 P (1500 Pa s) at temperatures of up to 900°F with line pressures of 670 psig (4.6 MPa). Since the viscosity depends on the temperature of the fluid, the measuring system must be temperature controlled using a resistance thermometer either to measure the temperature at which the viscosity is measured or to control the temperature in the measured fluid to maintain a constant viscosity. Overall error is ±1% with a response time of about 2 min.*

Several viscometers use a rotameter type of float. The single-float viscometer as shown in Figure 6-2a is a direct reading instrument for the continuous measurement of viscosity. A positive displacement pump provides a constant sample flow rate.

The single-float viscometer can be used with non-Newtonian fluids with viscosities less than 400 cP and Newtonian fluids up to 10,000 cP. The maximum span is 6:1 and the minimum is 3:1. Accuracy is ±4% of indication, and reproducibility is ±1%. The usable flow rate is 1 to 2 gpm (3.8 to 7.6 L/m).

The two-float viscometer is generally used for local indication and signal transmission. It employs two floats. The fluid flow rate is adjusted to the value indicated by the position of the upper float. By maintaining a constant reference flow rate as indicated by the flow rate float, the position of the other float indicates the viscosity of the fluid. The two-float viscometer can be used for Newtonian fluids with viscosities from 0.3 to 250 cP and with a span range of

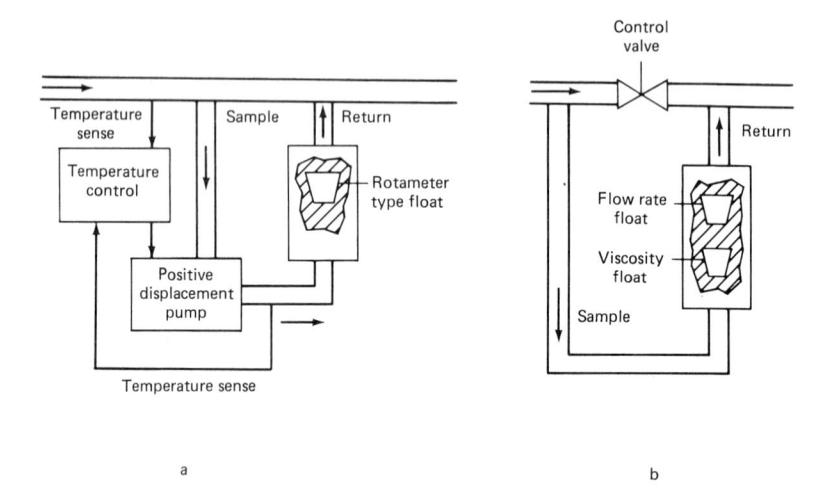

Figure 6-2. Float-type viscometers. *a.* Single float. *b.* Dual float.

---

*Some designs place the capillary tube directly in the process stream, which reduces the response time but increases the error to about ±2%.

10:1. The error is ±4% for viscosities higher than 25 cP, and ±2% for lower viscosities. The reproducibility is ±1%. The usable flow rate is 0.25 to 2.5 gpm (0.95 to 9.5 L/m).

The concentric viscometer uses a differential pressure regulator to maintain a constant pressure drop and a variable-area flowmeter with a viscosity-sensitive float. As the fluid enters the unit, it splits into two streams. The fluid that flows upward around the differential pressure float controls the pressure drop. The upper end of the differential pressure float is used as a control valve, and as the flow rate changes, it throttles the flow to maintain a constant pressure drop which is determined by the weight of the float. That portion of the fluid that flows downward enters the viscometer tube through an orifice and then passes the viscosity-sensitive float.

A constant flow rate through the tube is maintained as the fluid flows through the orifice at a steady pressure drop. Measurement of the viscosity is made under the constant flow rate. The float transmits its movement through magnetic coupling. The usable slow range is 1 to 10 gpm with pressure and temperature ratings to 650 psig (9.5 mPa) and 450°F (232°C).

OSCILLATION TECHNIQUES

The ultrasonic viscometer as shown in Figure 6-3 consists of a probe and associated circuitry connected by a coaxial cable. An electronic oscillator transmits pulses of current to a coil inside the probe and around the thin blade. The resulting field excites the magnetostrictive member and causes the blade to vibrate at its natural frequency, which is determined by the length of the strip. The probe acts as the transducer, measuring the damping effect of viscous drag of the liquid.

The damping effect on the ultrasonic vibration is a function of the liquid density as well as of the viscosity, so compensation is normally required. This viscometer is widely used in many chemical, petroleum, paper, textile, rubber, plastic, and paint industries.

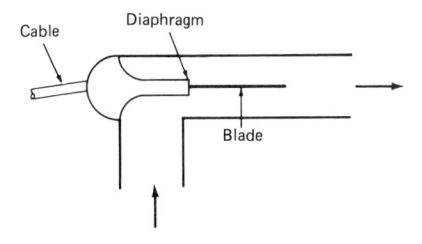

Figure 6-3. Ultrasonic viscometer.

132 TRANSDUCERS FOR AUTOMATION

The vibrating-reed-type viscometer as shown in Figure 6-4 consists of a frequency generator, vibrating spring rod, probe, and pickup unit. In operation it uses the amplitude of the probe vibration to determine the viscosity. The resistance to shearing action caused by the probe vibration increases with the medium viscosity. The amplitude of probe vibration is transferred through a welded node point at the pickup end of the detector. The pickup is similar to the drive system except it is used to induce a voltage in the pickup coil of coils. This voltage is usually 200–800 mV. In practice this viscometer is normally installed in a process loop where temperature, pressure, and flow rate are controlled in order to maintain laminar flow. Complete immersion of the probe is also required.

TORQUE AND WEIGHT TECHNIQUES

A rotating-cone viscometer is shown in Figure 6-5. It measures viscosity by sensing the torque required to rotate a spindle in a liquid.

In a process automation application, the sample is continuously replaced and is subjected to a constant shear rate. The measurement of the non-Newtonian apparent viscosity is possible, as well as of the absolute viscosity of Newtonian fluids.

A synchronous induction motor is used to drive a cage coupled through a calibrated spring to a spindle arm which supports the spindle or cylinder in the fluid being measured. To obtain a measurement, the spring tends to wind until its force equals the viscous drag on the spindle. The cage and spindle then rotate at the same speed but with an angular relationship to each other which is proportional to the torque on the spring.

In some units a variable capacitor is used. The final capacitance is proportional to the angular relationship between the cage and spindle. Another type of rotating-cone viscometer uses a potentiometer to sense the angular displacement.

This type of torque element viscometer is a flow-through system with the

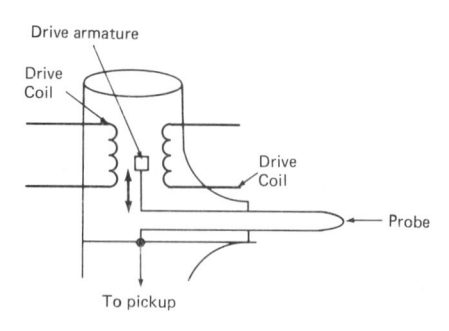

Figure 6-4. Vibrating-reed viscometer.

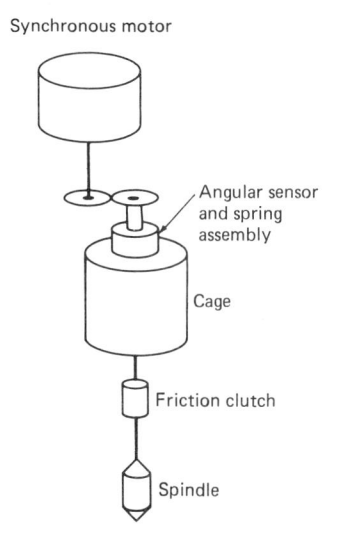

Figure 6-5. Rotating-cone viscometer.

flow upward or vertically into the measuring system, which is housed in a stainless-steel flow-through body. Operational range is 50,000 cP (50 Pa s) with an error of ±1% and a repeatability of ±0.3%. The response time is about 30 s for a full-scale change. The spindle speed is typically 50 rpm and changing the spindle size allows the viscosity range to be adjusted.

The total system consists of a housing; a measuring cell made up of a measuring cup; a magnetic coupler; and a torque meter made up of a three-speed gearbox, synchronous motor, capacitor or potentiometer, and torque spring.

The agitator power viscometer operates in a similar manner to the rotating-cone viscometer except that the torque exerted is measured by a transmitting wattmeter or thermal converter. It measures the power consumed in driving an agitator in the mixing tank. Different impeller designs are available for different ranges. This viscometer has been used in the paper industry to control and measure the consistency of paper pulp slurries. It tends to be self-cleaning and the agitating design makes it ideal for materials that have a tendency to cling to parts or to settle out from suspensions. It cannot be used with fluids which are thixotropic or rheopectic since these change viscosity under agitation.

The falling-piston viscometer is shown in Figure 6-6. A major feature of this design is measurement reproducibility, which allows its use for the in-process measurement of both Newtonian and non-Newtonian viscosities.

As the piston is raised, the sample of liquid that is to be measured is drawn in through openings in the sides of the tube. This liquid fills the tube as the piston is completely withdrawn. During the measurement, the piston is allowed

134 TRANSDUCERS FOR AUTOMATION

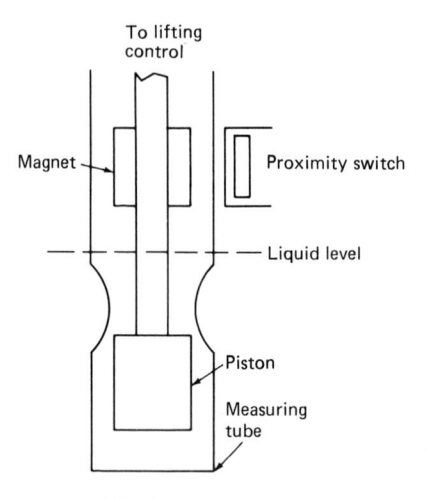

Figure 6-6. Falling-piston viscometer.

to fall by gravity, forcing the sample out the tube through the same route as it entered. The time of fall is a function of the viscosity, which also depends on the clearance between the piston and the wall of the tube since this acts as an orifice. A variation on this type of viscometer uses a cylindrical slug with magnetic switches in a fall tube. The slug is allowed to rise to the top of the tube from the sample flow. Then the sample flow is cut off and the time of fall is determined from the actuation of the two magnetic switches. The measured time interval is calibrated in units of viscosity. This type of viscometer should not be used in automation applications when response times must be less than one minute.

The automatic efflux-cup viscometer is shown in Figure 6-7. These are low-

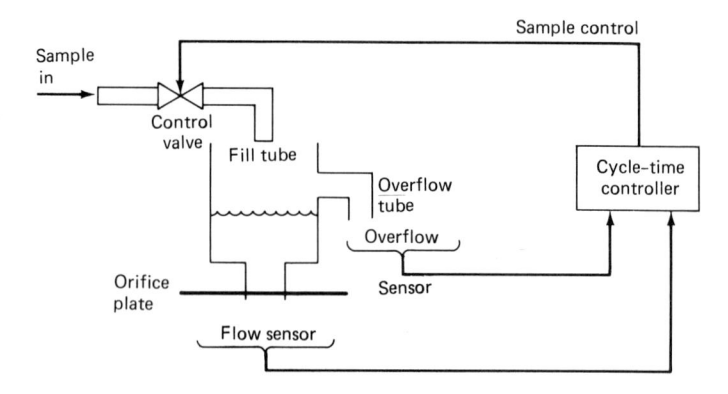

Figure 6-7. Automatic efflux-cup viscometer.

cost on-line devices that can be used where accuracy is not critical and intermittent measurements with major time lags between measurements are acceptable. The sensors are used to detect the efflux time, which is a function of the viscosity.

Fillings, efflux timing, and solvent washing operations are controlled by the cycle timer. Under the best conditions the liquid flows through the calibrated orifice in the bottom of the cup in 20 to 40 s. A repeatability of ±2% is possible under these conditions.

The selection of a standard viscometer for the application is not always possible and the unit may have to be customized in some way. Viscometer vendors or consultants can assist in selection and modification. Many plants that require on-line viscosity measurement as a major method of controlling process operations use customized devices.

MOISTURE AND HUMIDITY MEASUREMENTS

The moisture content of the atmosphere is vital in the automation of many industrial processes, such as the manufacture of textiles, tobacco, paper, soap, chemical solvents, fertilizers, and wood products. Humidity measurement and control are necessary during processes like the drying of chemicals. (See Table 6-1).

Psychrometry is the study of the dynamics of the atmosphere using the measurement of a number of variables. A psychrometric chart relates the basic humidity parameters.

One of these variables is the relative humidity, which is a measure of the water vapor present in the air. Relative humidity is defined as either (1) the ratio of the moisture content of the air to its saturated moisture content at the same temperature, or (2) the ratio of actual water vapor pressure in the air to the water vapor pressure in saturated air at the same temperature.

The dew point is the temperature at which the water vapor present in the air saturates the air and begins to condense as dew begins to form. When the temperature of the air is reduced below the dew point, the air remains saturated and the partial pressure of the water vapor decreases because of condensation.

Many humidity sensing elements also perform transduction. An example is found in humidity sensing resistive elements which are actually combination sensing/transduction elements.

The elements for humidity and moisture transducers can be grouped into three categories based on their measuring techniques.

1. Hygrometers measure humidity directly, with the resistive hygrometer being widely used. The use of older displacement producing techniques has been decreasing in favor of resistive elements.

TABLE 6.1. Humidity measurement methods.

Parameter	Description	Units	Typical applications
Wet bulb temperature (psychrometer)	Minimum temperature reached by a wetted thermometer in an airstream	°F or °C	High-temperature dryers, air conditioning, meteorology, controlled chambers
Percent relative humidity	Ratio of actual vapor pressure to saturation vapor pressure, with respect to water, at the prevailing dry bulb temperature	0–100%	Monitoring conditioned rooms, test chambers, pharmaceutical, and food processing
Dew/frost point	Dew point is the temperature to which the air must be cooled to achieve saturation; if the temperature is below 32°F, it is called the frost point		Heat treating, annealing, dryer control, air monitoring
Volume or mass ratio	Parts per million (ppm) by volume is the ratio of partial pressure of water vapor to partial pressure of dry carrier gas; ppm by weight is identical to ppm	ppm ppm	Used primarily to ensure dryness of industrial process gases such as air

2. Psychrometers measure humidity indirectly. Resistive temperature transducers are frequently used for "wet bulb" and "dry bulb" measurements.
3. Dew-point elements give a direct indication of dew point from which the humidity can be derived. In this group the cold mirror devices are the most common.

HYGROMETRIC TECHNIQUES

The use of resistive hygrometer elements is a popular method of humidity sensing. A change in relative humidity (RH) will produce a change in the resistance of a material such as a hygroscopic salt or carbon powder. The material is applied as a film over an insulating substrate and terminated with metal electrodes as shown in Figure 6-8. The hygroscopic salt film that results has a resistance that varies as a function of the vapor pressure of water in the air. A common hygroscopic salt is lithium chloride.

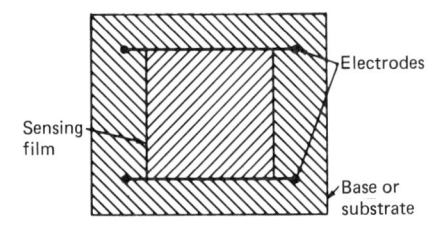

Figure 6-8. Typical resistive humidity sensing element.

A typical element uses a film consisting of an aqueous solution of less than 5% of lithium chloride in a plastic binder. The lithium chloride element is sometimes referred to as the "Dunmore element" or "Dunmore hygrometer." The Pope cell uses a similar construction with a polystyrene substrate in a sulfuric acid solution.

In the Dunmore type of hygrometer, there is a steep resistance to the relative humidity change that occurs. This makes it necessary to vary the element spacing or resistance properties of the film for specific humidity ranges. The result is that several different resistance elements are needed to cover a standard range. The Pope sensor has a wider range and allows a 15–99% range in a single element.

Systematic calibration is essential since the resistance grid varies with time and contamination from water cleaning sprays as well as with exposure to temperature and humidity extremes. Direct current generally polarizes these sensors so ac excitation is normally used.

A number of other materials are used. These resistance sensors include carbon strips, the aluminum oxide hygrometer, electrolytic conductive elements (phosphorus pentoxide), ceramic elements, poly(vinyl chloride) elements, silicone polymers, and certain types of crystals.

The aluminum oxide hygrometer element uses the electrical properties of anodized aluminum for humidity measurement in the form of small strip, needle, or rod elements. As the aluminum surface is being anodized, a thin layer of aluminum oxide is formed.

A thin metal coating of aluminum or gold is then deposited on the outside surface of the aluminum oxide layer. This acts as one electrode and the aluminum base acts as the other electrode, as illustrated in Figure 6-9a.

The number of water molecules absorbed on the oxide structure is a function of the change in impedance in the equivalent circuit of the hygrometer structure. The transduction is both capacitive and resistive. This impedance change is measured to indicate the humidity. A thin-film strip aluminum oxide element equivalent circuit is shown in Figure 6-9b.

A hygroscopic film is employed in some capacitive humidity sensing elements where it acts as a humidity-sensitive dielectric. Changes in the dielectric cause

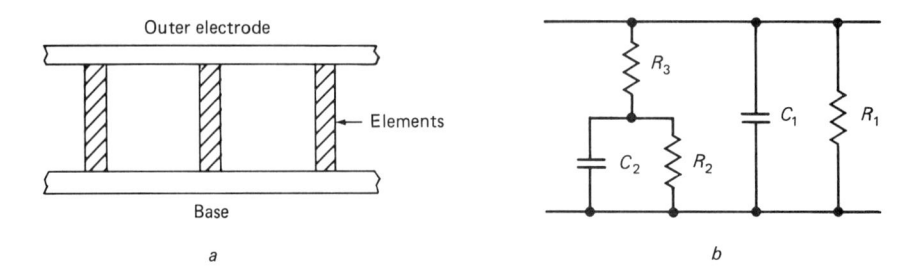

Figure 6-9. *a.* Aluminum oxide humidity sensor. *b.* Circuit of a single section.

the capacitance to change with the humidity. The capacitance of the element changes with the amount of water vapor in the dielectric.

Sensors of this type use a thin-film amorphous polymer capacitor on a glass substrate and exhibit the following typical specifications:

Range—0 to 100% RH
Response time—1 s to 90% humidity change at 20°C
Hysteresis—1% for humidity excursion of 5 to 80 to 0%, 2% for humidity excursion of 0 to 100 to 0%
Temperature coefficient—0.05% RH/°C
Drift (8 h)—1% from 0 to 60% RH, 2% at 80% RH, 5% above 90% RH
Accuracy—±3% from 5 to 80% RH, ±5 to 6% from 80 to 100% RH
Sensitivity—0.2% RH
Input power—3.5 ±0.01 V dc
Power consumption—10 mA
Output—0 to 100 mV
Operating temperature range −40 to 120°F (−40° to 50°C)

Oscillating-crystal hygrometers measure humidity with a quartz crystal that has a hygroscopic coating. The crystal operates in an oscillator circuit. The mass of the crystal changes with the amount of water in and on the coating, which changes the frequency.

This type of hygrometer may compare the changes in frequency of two hygroscopically coated quartz-crystal oscillators. As the mass of the measuring crystal changes as a result of the absorption of water vapor, the circuit frequency changes. Thus, the oscillations reading is related to the moisture content of the incoming air. Most commercial units of this type use an internal system to produce a dry reference gas and the frequency change is corrected for flow and the type of gas. Many units use lithium chloride as the hygroscopic coating on the crystal. The operating frequency is about 10 MHz. When two crystals are used, while one is in service the other is being dried by the reference gas system.

This type of instrument is relatively expensive. Its flow sensitivity, susceptibility to damage by contact with water, and calibration dependence on a carrier gas have made it a difficult unit to employ in process automation applications.

A spectroscopic hygrometer measures humidity based on the analysis of absorption bands of water vapor in a gas sample. These units employ a sensing path as the sensing element, usually in the form of a cylinder, through which a beam of radiation passes. This beam may be infrared, microwave, ultraviolet, or visible. The absorption characteristics are determined in the gas sample in the sensing path.

This type of hygrometer requires an energy source, an energy detector, an optical system for isolating wavelengths in the spectral region of interest, and a measurement system for determining the attenuation of radiant energy caused by the water vapor present in the optical path. In some units a sample cell contains the gas being sampled and provides the means for calibration. Separate calibration gas bottles may be used to provide zero and span. Sampling lines, flowmeters, and pressure regulators are also required in these systems.

A mechanical hygrometer uses a material which changes dimension with the absorption and desorption of water from the air. Organic materials have been used for humidity sensing elements longer than any others. The first and most popular material was hair, particularly human. The second was animal membrane. Other materials, such as paper, wood, textiles, and plastics have been used but they have proved to be less satisfactory due to aging and hysteresis.

Human hair and animal membrane change length with increasing humidity. This dimensional change has been used in a number of different designs. The change in length of human hair over the range of 0% to 100% RH tends to be small and nonlinear. The sensitivity decreases from approximately 0.0005 in. of hair length per % RH to about 0.00005 in. per % RH at 85% when a one-gram tensile force is applied. The material has a large hysteresis effect and measurements are generally in error below 32°F. The time response is usually too large for monitoring a changing process. Strain gauge sensors are used in a similar way.*

PSYCHROMETRIC TECHNIQUES

Psychrometric elements measure humidity utilizing the "wet and dry bulb" method of temperature sensing. The two psychrometric elements have separate outputs. One of these elements, the dry bulb, measures the temperature of the ambient air. The other element is the wet bulb, which is enclosed by a wick that is saturated with distilled water. As the measurement is made, the ambient

---

*The strain gauge sensors have similar performance characteristics to the hair type of sensor but they tend to be more rugged and not as easily damaged by water spray.

air ventilates the wick and cools the sensing element below the ambient temperature due to the evaporation of water from the wick. The evaporation of moisture from the wick is proportional to the vapor pressure or moisture content of the ambient air.

The actual temperature sensing elements are usually resistive. The wick may be a textile material or a porous ceramic sleeve which is fitted over the resistive element.

This type of steady-state, open-loop, nonequilibrium process is dependent on the purity of the water; the cleanliness of the wick; heat radiation effects; the accuracy of the temperature sensor; and the density, viscosity, and thermal conductivity of the gas. These error sources are also functions of the pressure, temperature, and type of gas.

Most systems and the corrections that are used are valid for wet bulb temperatures above 32°F (0°C). If the bulb becomes encrusted with ice, special correction factors must be used. When properly used with atmospheric pressures and gases, this technique is a useful calibration standard. In a process, it is susceptible to operator error and contamination problems.

DEW-POINT SENSING TECHNIQUES

Dew-point sensing elements measure a discrete temperature. This is the temperature at which liquid water and water vapor, or ice and water vapor, are in equilibrium. At this temperature only one value of saturation vapor pressure for water vapor exists. The absolute humidity can be determined from this temperature and a knowledge of the pressure.

To find the dew point at any given air or gas temperature, the temperature of a surface is lowered until dew (or frost) first condenses on it. As this point is reached, the temperature of the surface is measured.

The dew-point sensing element must perform the function of temperature sensing as well as sensing the change from vapor to liquid (or solid) phase. The temperature of the surface must be measured at the instant when condensation first occurs since the characteristics of the condensate will not change appreciably as the surface continues to be cooled below the dew point.

Dew-point sensing elements sense relative humidity as do psychrometric sensing elements. In both cases, tables are used from which the relative humidity values are calculated from the measurement obtained and the ambient temperature. The relative humidity is determined from the dew-point data by the use of these saturation vapor-pressure tables. These tables can be stored in semiconductor memory and used by a microcomputer.

The time of condensation function is usually provided by a thin disk or plate with a smooth surface. This disk or plate is closely coupled thermally to a cooling element and a condensation detector. Thermoelectric coolers are most popular.

The various condensation detectors include

1. Photoelectric types which use a mirror as the condensation surface, a light source to illuminate the surface, and one or more light sensors which detect light reflections from the surface.
2. Resistive units which use a metal inlaid in the condensation surface; here a change in the surface resistance occurs when the condensation forms.
3. Nucleonic units in which an alpha or beta particle radiation source is located within the condensation surface and a radiation detector senses the drop in radiation when condensation forms over the source.

The basic operation of these condensation detectors is shown in Figure 6-10.

The temperature at which condensation first occurs upon cooling of the surface is usually sensed by resistive or thermoelectric sensing elements. Platinum-wire resistance temperature sensors are used more often than thermistors or thermocouples but all three types have been employed in dew-point units. Typical accuracy is $\pm 1°F$ with special designs capable of $\pm 0.05°F$.

An actual thermoelectrically cooled, optical dew-point hygrometer uses a continuous sample of the atmosphere gas over the mirror. The mirror is illuminated by a light source, and observed by a photodetector bridge network. The change in reflectance is detected by a reduction in the reflected light due to the light scattering effect of the individual dew molecules. The reduction in light forces the optical bridge toward a balance point, reducing the input error signal to an amplifier which controls the power supply to the thermoelectric cooler. In this way the mirror is held at a temperature which retains a constant-thickness dew layer. Embedded within the mirror, a temperature sensor measures the dew-point temperature.

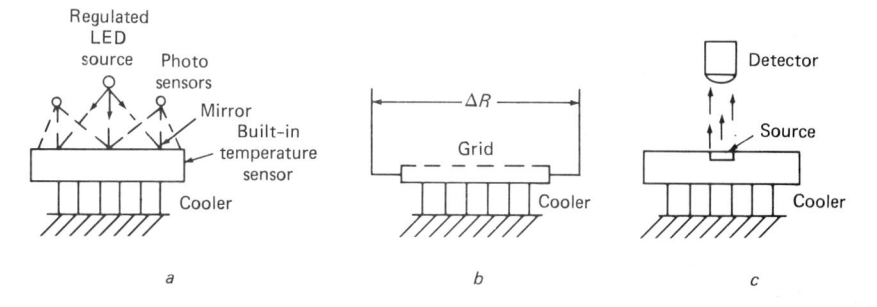

Figure 6-10. Dew-point sensing techniques. *a.* Photoelectric. *b.* Resistive. *c.* Nuclear.

This type of hygrometer is continuous measuring, rugged, direct reading, and suited for process automation. It is relatively expensive because of the design and the accuracy of measurement obtained. Most types can be used on clean gases down to -40 to $-100°F$.

Another design uses a lithium chloride element as a self-regulating heater. The element heats itself from the vapor pressure of a lithium chloride solution which is in equilibrium with the pressure of the ambient atmosphere. A resistive or thermoelectric temperature sensing element measures the temperature of the solution. Humidity sensor characteristics are summarized in Table 6-2.

HUMIDITY MEASUREMENT ERRORS

An important factor that must be considered in humidity measurement is the sampling system, which can be a probe sample cell (Fig. 6-11). The leakage, pressure, and temperature gradients and moisture absorption/desorption characteristics of the sampling system can all be potential error sources.

The errors due to leakage will depend on the ambient conditions. If the dew point being measured is close to the ambient dew point, leakage into the system may not change the reading substantially. If the system is pressurized above atmospheric so as to create a leakage out of the system, the error introduced will be even less.

The temperature stability of the sampling system components can even be more important. For a given equilibrium sampling condition a specific amount of moisture will be absorbed by the sampling system's wetted surface. Any factors which upset this equilibrium, such as a change in sample concentration of the process or an ambient temperature fluctuation of the sampling system, will cause a new equilibrium condition to be established before the true dew point can be measured. Thus, any control of the sample line temperatures should be noncyclic; a constant or proportional control should be used instead of an on-off control. A portable calibration system like that shown in Figure 6-12 can be used when equilibrium factors may change.

The effect of absolute pressure on the measurement is also important. Dew-point analyzers determine the dew point at the actual total pressure within the sensor. If a mixture of gas and water vapor is subject to a change in pressure and there is no precipitation or addition of water to the system, Dalton's law states that the partial pressure of the water vapor must change in the same ratio as the total pressure. This causes a change in the dew point for a change in total pressure.

Errors can also be caused by the material absorption/desorption characteristics on the overall system response. Stainless-steel, glass, and nickel alloy tubing tend to be the best nonhygroscopic materials for low-dew-point applications ($0°F$ to $-100°F$). Teflon begins reducing the system response due to desorption

TABLE 6.2. Humidity sensors.

Type	Advantages	Disadvantages
Relative humidity	Useful for small areas, fast response, good accuracy	Secondary measurement, subject to hysteresis and aging, narrow operating range, condensation possible near 100% RH, accuracy limited below 10% RH
Mechanical relative humidity	Low cost, low sensitivity to contaminants	Limited accuracy, subject to aging and hysteresis
Wet/dry bulb psychrometer	High accuracy near 100% RH, low cost, usable to 212°F	Low accuracy below 20% RH, lower limit 32°F (0°C), adds moisture to area being measured, must be serviced often
Aluminum oxide moisture	Small size, multiple sensors available, temperature range to below −100°F	Subject to aging and hysteresis, limited accuracy, secondary device, sensitive to contaminants, expensive
Electrolytic cell	Basic measurement, more accurate than Al_2O_3, good repeatability, good stability, water-vapor specific	Controlled sample flow required, sensitive to contaminants, not available in multiple sensors, 2000 ppm maximum, not easily serviced, not usable with all gases
Chilled-mirror dew point	High accuracy, NBS traceable, broad dew point and temperature range, easily serviced, fast response, rugged, reliable,	Not water-vapor specific, high cost, cannot be used with ammonia, chlorides, or SO_2, slow response at low frost points, subject to contaminants
Lithium chloride dew point	Simple construction, low cost, very reliable, basic measurement, not sensitive to contaminants	Slow response, narrow range, not as easily serviced, not usable below 12% RH, adds heat to measured area

at the lower dew points. Copper, aluminum alloys, and stabilized polypropylene tubing are useful above the −20°F dew point.

Most plastic and rubber tubing is unacceptable in all ranges. Unless it is attacked by the sample, the effect of these more hygroscopic materials is not that of contamination, but of introducing a severe time lag into the system during

144 TRANSDUCERS FOR AUTOMATION

(a)

Figure 6-11. *a.* A moisture probe sample cell such as this provides a simple bypass loop which can be adequate for many measurements. The bypass allows servicing of the probe without affecting most measurements. *b.* This aluminum oxide sensor is composed of three layers. The middle layer is the aluminum oxide porous surface, which is covered with a thin layer of gold and mounted on an aluminum base. *c.* The sensor is then covered with a porous protective cone to complete the probe assembly, which is then inserted in the sample cell. (*Courtesy Panametrics*)

the establishment of an equilibrium condition. Plastics such as nylon cannot be used at low dew points since it can take days to reach the equilibrium condition.

System contamination effects on dew-point measurement can be caused by condensibles or noncondensibles. When the optical dew-point analyzer measures the dew point, any substance that condenses on the mirror surface may contain contamination on constituents which will not condense unless their dew-point temperature is reached. These condensibles may be soluble or insoluble. They may be insoluble with a dew point at or above that of the constituent being

(b)

Figure 6-11 (*Continued*)

measured; then the relative concentration level will determine the effect on the measured dew point.

When the concentration level of the contaminant is low, it will have a low partial pressure compared to the water vapor. It can then be removed by heating the mirror surface.

At high concentration levels the optical dew-point analyzer may measure the dew point of the contaminant rather than the water-vapor dew point. This problem is reduced as a result of the high light attenuation characteristics of dew or frost compared to many common contaminants. If the contaminant is soluble in the constituent being measured, it will modify the vapor pressure and, thus, the dew point of the degree of solubility.

(c)

Figure 6-11 (Continued)

HUMIDITY AND MOISTURE AUTOMATION CONSIDERATIONS

Dry and wet bulb temperatures can be used to estimate relative or absolute humidity or dew point. The relative humidity and dry bulb temperature can be used to determine the absolute humidity and therefore the dew point. In the automation of a process, the proper choice of a measurement technique which best represents the state of the process can be critical.

When air is to be held at one point on the psychrometric chart, two variables must be controlled. In order to minimize the interactions between these two control variables, the choice of measurement technique must relate closely to the manipulated variable.

For example, we could control air conditions using both a heating coil and a water spray. The heat will affect the dry bulb and wet bulb temperatures, as well as the relative humidity, but it will not change the dew point. The water sprayed into the air lowers the dry bulb temperature and raises the relative and absolute humidities as well as the dew point, but the wet bulb temperature will remain steady. An optimal control scheme might have a dew-point controller

VISCOSITY MEASUREMENT 147

Figure 6-12. A portable calibration system such as this can generate a repeatable concentration of water vapor in a carrier gas stream. It is based on saturation/dilution techniques. (*Courtesy Panametrics*)

for the water flow, and a wet bulb controller for adjusting the heater. This particular combination of loops will tend not to interact. We might also apply other controlled variables such as steam injection or a fresh air input and other measurement choices would be used.

If it is required to provide air of a certain moisture content at a constant barometer reading but variable temperature, dew-point control can be used. Variations in pressure must be considered, and if the operating pressure is variable a composition analyzer may be needed to determine the partial pressures.

In many applications there can be problems from condensation. To prevent these corrosive conditions, we should control the temperature difference between the dew-point and the dry bulb measurements.

Materials like paper and wood will change their dimensions with relative humidity. To control the dimensional stability of these materials one can measure the dimension of a similar material like hair. The hair hygrometer is a useful humidity measuring device for this type of application.

Other problems are present in the drying of solid materials. When air is heated at constant humidity and then blown across the moist material, adiabatic evaporation occurs. As a result, the air's wet bulb temperature is the same before and after contact with the product material. Usually the temperature of the material is similar to the wet bulb temperature of the air, except if extreme dryness is present,. The wet bulb wick thus acts as a model of the solid being dried. The evaporation rate is proportional to the temperature difference between the air and the product and this is the difference between the dry and wet bulb temperatures. Thus, we conclude that relative humidity and dew point are not useful variables for dryer control.

EXERCISES

1. Define a poise with a labeled diagram.
2. Discuss the differences between Newtonian and non-Newtonian substances.
3. Diagram the apparatus required for some basic methods of determining the viscosity of oil.
4. Discuss how the viscosity of liquids and gases can change with respect to temperature and pressure changes.
5. Diagram and discuss the following methods of determining the viscosity of a liquid: (a) continuous capillary method, (b) float method, (c) vibrating-reed method, (d) torque method.
6. Prove that the following expression is true: Kinematic viscosity (centistokes) = absolute viscosity (centipoise)/Mass density, provided that the temperature recorded at the viscosity reading does not change.
7. A solid sphere of radius r cm sinks in a viscous liquid at a constant velocity of v cm/s and u is the absolute viscosity of the liquid, in poise. The resistance to the motion of the sphere is found to be

$$R = 6urv \text{ dynes.}$$

Compute the viscosity of the liquid in which a sphere of diameter 0.0645 in. sinks 20.5 cm in 20.6 s. The density of the liquid is 0.76 g/cm^3 and that of the sphere is 7.4 g/cm^3.

8. An oil with a kinematic viscosity of 0.006 ft/s flows through a 2.0-in.-i.d. pipe with a velocity of 1.5 ft/s. If the oil weighs 58 lb/ft, calculate the friction drop, in pounds per square inch, over 1000 ft of pipe.
9. Describe the operation of a continuous-reading viscometer monitored by a microcomputer control. Draw a flow diagram for the microcomputer program.
10. Discuss the importance and relationship among the following terms in dimensional control: (a) dew point, (b) relative humidity, (c) specific humidity, (d) partial water vapor pressure, and (e) dry and wet bulb temperatures.
11. Describe the difference between hygrometers and psychrometers.
12 Consider the operation of an aluminum oxide hygrometer. How would you interface it with a microcomputer in an automated system?
13. Describe how an oscillating-crystal hygrometer would be connected as a control input to a microcomputer system. Draw a flow diagram for the interface.
14. What are the advantages and disadvantages of the spectroscopic hygrometer?
15. How can the properties of mechanical hygrometers be used in industrial automation?
16. Describe the operation of the various types of condensation detectors available for optical dew-point sensors.
17. Why is it important to know both the process pressure and the pressure in the dew-point sensor? Is this true of all gas analysis instrumentation?
18. How does the area of leakage affect humidity measurement in systems operating below ambient pressure?
19. Consider the effects on dew-point measurement of noncondensible contaminants which are mainly salts. How could a microcomputer be used to reduce or minimize these effects in an automated system?
20. Describe how one could use a combination of any two measurements of humidity and temperature to control the condition of air in a drying application. Draw a flow chart for a microcomputer control which could be used to optimize the drying system.

7
LEVEL MEASUREMENTS

APPLICATIONS

Level measurement applications can be grouped as follows: atmospheric vessels, pressurized vessels, and accounting grade. Accounting-grade measurements are made in both atmospheric and pressurized vessels.

Liquid level detection in atmospheric vessels rarely presents major problems and a number of desirable features can be built into an atmospheric vessel level system. The instrumentation generally can be selected and installed so that it can be removed from the vessel for calibration or repair without draining the vessel.

Solid level gauging is generally done in atmospheric tanks, but there are fewer available level detecting devices and less flexibility in installation methods. The devices that are suitable for point level detection of solids range on the following scale:

Radiation   Ultrasonic   Rotating   Diaphragm   Capacitance   Conductivity
                         paddle
            Vibrating   Antenna   Impedance
            reed

Most Suitable ——————————————————————————————— Least Suitable

Point detection units must be located at the actuation point, and this can cause accessibility problems. Except for the radiation type, it also requires that a new connection be cut into the vessel if the actuation point is changed.

Solids that can behave unpredictably can cause measurement problems. If the solid is viscous, particular care must be taken in the location and installation of the detector. Continuous level measurement of solids can be made by using the proper units.

The surface sensor design is the most often used. All of these designs require top mounting, but since they can be equipped with ground or remote readouts, this is not a serious limitation. However, the performance of these instruments is not good when the solids are not free flowing. If an inventory or accounting-grade measurement is required, it may be preferable to make a weight measurement with load cells.

The point level detection of liquids in a pressurized vessel can be made using point detecting switches. For clean applications in industrial processing, one can use an externally mounted displacer switch. This type of device is rugged and reliable, it has above average resistance to vibration, and some designs allow the actuation point to be changed easily.

Capacitance switches or float switches may be also be used if they can be arranged so that they may be removed for repairs without depressurizing the vessel. Conductivity switches can be used in water services to 700°F (370°C) and 3000 psig (21 MPa). They should not be automatically considered for hydrocarbon or chemical services. The balance of the switches are less expensive and are generally used in noncritical applications and atmospheric vessels.

The requirement for accuracy in accounting-grade installations may be demonstrated as follows. If a 750,000-barrel storage tank has a diameter of 345 ft (105 m) and it takes some 8000 gallons to raise the level 1 in. (25 mm), a level measurement error of 1 in. could indicate that 8000 gallons (30 kL) have been gained or lost. This is a matter of concern, especially if the level measurement is used as a basis for custody transfer of the material.

ANTENNA LEVEL SENSORS

In an oscillator circuit, the output is fed back to the input in such a way that the loop gain is greater than one and the input is in phase with the output. The circuitry is self-limiting in that a gain of greater than one does not cause control problems.

The frequency of oscillation of the circuit can be made to vary if passive elements in the circuit are varied. For example, a variable capacitor can be employed in the circuit in such a way as to make the oscillating frequency proportional to the varying capacitance. If the capacitance is proportional to the level in a tank, the oscillation frequency will also be proportional to this level. The antenna level sensor uses this principle.

When an antenna is designed to be installed in air, the dielectric constant of the surrounding material is assumed to be 1.0. If an antenna is installed in a bin that has a rising and falling process material level, the value will be different because the material will have a dielectric different from that of air.

The antenna to bin wall capacitance will vary with the changing level. By connecting the antenna to an oscillator circuit, the frequency of oscillation can be made to vary with the level. The frequency of oscillation of the circuit containing the antenna can be compared with the frequency of oscillation of a constant-frequency oscillator and the difference used for point level or continuous level measurement.

In a typical installation of an antenna probe for point detection, the antenna is installed horizontally (Figure 7-1a) so that a large portion of the antenna is covered or uncovered at once, causing a large change in capacitance with a large change in the frequency of oscillation.

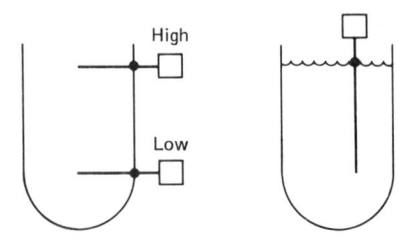

Figure 7-1. Antenna level sensors. *a*. Point detection. *b*. Continuous detection.

The continuous detector is suspended from the top of the bin and hangs down into it as shown in Figure 7-1*b*. The capacitance changes linearly with increasing level. The antenna may be a solid rod or stranded cable. It can be bare or coated with Teflon or other plastics. Insulative coatings are used when the materials are electrically conductive.

Antenna probes have several disadvantages and limitations:

1. Changes in the dielectric constant of the process material can affect the accuracy.
2. Radio frequency devices can interfere with proper operation.
3. The bin wall must be grounded, so fiberglass bins cannot be used.
4. Conducting and nonconducting buildup on the probe can affect accuracy.

Another problem is that the system will always contain inherent nonlinearities. The capacitance can be made to change linearly with level, but the oscillation frequency will not change linearly with the varying capacitance. This means that the antenna/oscillator circuit can only be calibrated for accurate performance at one point (one capacitive value) and that the performance of the system will become increasingly nonlinear as the system operates farther away from that point.

BUBBLER SYSTEMS

The bubbler system is used to measure liquid level with compressed air. A dip tube is installed in a tank with the open end about 3 in. (76 mm) from the bottom. As air is forced through the tube, air bubbles escape from the open end and the air pressure in the tube equals the hydrostatic head of the liquid. As the liquid head varies, the air pressure in the dip tube changes and this pressure can be detected to indicate the level.

Figure 7-2 shows a typical air bubbler installation for an open (atmospheric) tank. The top of the dip tube may be cut to guarantee the continuous flow of small bubbles. The transmission line should be sloped toward the tank so that,

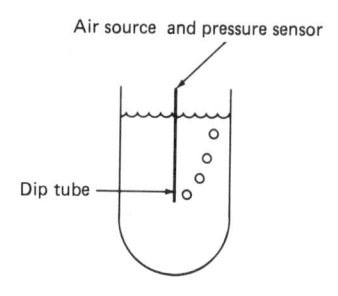

Figure 7-2. Bubbler system.

if process vapors enter the transmission tube, the condensate will drain back into the vessel. If the pressure readout device must be below the tank level, a condensate trap can be installed as shown.

The dip tube is fabricated from $\frac{1}{4}$- or $\frac{1}{2}$-in. (6- or 12-mm) diameter tube of a material compatible with the process. Plastic and metal tubing runs longer than 6 ft (1.8 m) as well as tubing in agitated tanks require continuous support. If the tank is agitated, or if the installation of a top-entry dip tube would interfere with other equipment, the tube could be installed in the side of the vessel below the expected minimum level of the liquid.

The purge supply pressure should be at least 10 psi (69 kPa) higher than the highest hydrostatic pressure to be gauged. The purge flow rate is usually kept small, about 1 scfh (500 cm^3/min), so that there is no significant pressure drop in the dip tube. The purge medium is usually air or inert gas, although liquids may also be used.

Several methods of gas purge control can be used depending on the installation considerations. A nitrogen supply pressure can be regulated to a value corresponding to a higher pressure than the hydrostatic head with the tank full. Purge flow may be adjusted by a needle valve. A rotameter or a sight feed bubbler can be used to adjust the purge rate; a rotameter can withstand higher pressures.

In this type of system, as the liquid level varies, the downstream pressure will also vary, causing variations in the purge flow rate. Since the purge pressure at the readout device is the sum of hydrostatic head and dynamic dip tube pressure drop, variations in purge flow will cause nonlinearities. To correct this condition, a differential pressure control valve can be installed across the fixed restriction of the needle valve. This will cause the purge to flow at a uniform rate regardless of the liquid head.

If the process material can build up or plug the dip tubed, an aerator line can be installed to allow for perioding blowing out of the transmission line.

In remote locations where gas purge media are not available, water may be jetted across a gap while air is aspirated into the stream and compressed. This air/water mixture enters the dip tube and the small amount of water runs down the inside of the bubbler tube while the pressure of the escaping air is detected as a measure of level. This would be in service only when the operator wanted to make a level reading, so that the water would not flow into the vessel continuously. A simpler approach for bubbler installations in remote locations is to use a small hand pump to compress the purge air.

For tanks that operate under pressure or vacuum, the installation of a bubbler system becomes more complex because the liquid level measurement is a function of the difference between the purge gas pressure and the vapor pressure above the liquid. A differential measurement is taken with all of the previously discussed variations applying.

Bubbler systems are not common in industrial process applications because of their limited accuracy and the fact that they introduce foreign matter into the process. Liquid purges may upset the material balance of the process and gas purges may overload the vent system in vacuum processes. If the purge medium fails, not only is the measurement lost, but the system is exposed to the process material, which can cause plugging, corrosion, freezing, or safety hazards.

Bubbler systems do allow economical gauging in applications such as waste water handling, food processing, and some bulk storage areas.

CAPACITANCE PROBES

A capacitance probe consists of two conductors isolated by an insulator. The conductors become the plates and the insulator is the dielectric. The characteristic feature of a capacitor is its ability to accept and store an electric charge. The larger the capacitor, the more current will flow to charge the unit. The size of a capacitor is affected by its physical dimensions and by the material between the plates; the dielectric.

For pure substances, the dielectric constant is a basic property. The relationship between the dielectric constant of a binary mixture and the percentage of one of the ingredients can best be established experimentally.

A change in the characteristics of the material between the plates will cause a change in dielectric constant, which is often larger, more definite, or more easily measured than changes in other properties. This makes the dielectric measurement suitable for the detection not only of level, but also of composition, moisture content, or chemical structure of a substance. Because the dielectric constant of gases is nearly unity, gas consumption cannot be measured by capacitance techniques.

Changes in process material will change the dielectric constant. While these changes are helpful in measuring the various characteristics of the material, they also affect the accuracy of level measurement.

The ideal capacitor is defined with infinite parallel plates. Since this situation cannot exist in the physical world, the capacitance will be a function of the geometry of the plates and will not perform linearly at the ends.

As the level of a vessel rises, the vapors with a low dielectric constant are displaced by the higher dielectric process materials and the capacitance changes can be detected with an instrument calibrated in units of level.

Chemical and physical composition and structure changes will affect the dielectric constant. The dielectric constant of solids is affected by variations in average particle size and changes in packing density.

As the material temperature increases, its dielectric constant tends to decrease. Temperature coefficients are usually on the order of 0.1% per degree Celsius. Automatic temperature compensator units are sometimes used to cancel the effect of temperature variations.

Variations in process level can be measured by a bridge circuit excited by a high-frequency oscillator (0.5–1.5 MHz.). As shown in Figure 7-3, the probe is insulated from the vessel and forms one plate of the capacitor and the vessel forms the other. The process pressures and temperatures determine the type of seal used at the insulator and the corrosion conditions determine the type of probe material.

For the measurement of nonconductive materials, a bare metal probe can be used. The effective resistance between the probe and vessel varies with the level in the vessel, and if it is not high enough to be approximated as infinite, the measurement cannot be made with a bare probe. To measure the level of conductive materials, an insulated probe must be used.

If the process material adheres to the probe, a level reduction in the vessel will leave a layer of fluid on the probe. If this layer is conductive, the reading will indicate the level to which the probe is coated. If the probe coating is nonconductive, the error is much less pronounced.

Another problem encountered in capacitance probe installations is ground paths through the head assembly. Moisture can enter through capillary leakage or a cable entrance, causing a ground path.

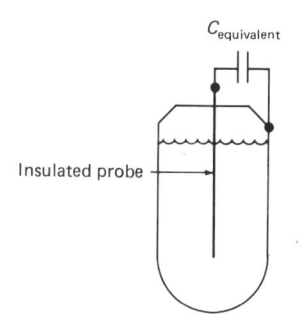

Figure 7-3. Capacitance probe.

Sensitivity and drift can also be a problem. Some devices are sensitive to 0.5 pF and will drift 5 pF for a temperature change of 100°F (56°C). The better units are sensitive to 0.1 pF and have a drift of 0.2 pF per 100°F (56°C).

Capacitance measurements can also be made in pipelines. Petroleum products can be distinguished by their different dielectric constants. By passing the pipeline product through a capacitive element, interface arrivals can be detected. These in-line dielectric detectors can also be used to measure other process properties that vary with dielectric, such as moisture content.

Capacitance probes can also be used for flow measurement across weirs and flumes. The flow across weirs and flumes is an exponential function of the upstream head. A capacitance probe can be characterized such that the capacitance changes with the head level are linearly proportional to the flow rate. One method of doing this is by characterizing the effective plate area.

The advantages of capacitance level measurement include simplicity of design, absence of moving parts, and proximity design, which requires no contact with the process. The disadvantages are that accuracy is affected by changes in the dielectric and process buildup on the probe. Because of these drawbacks, and recognizing that capacitance measuring installations are relatively expensive, there should be a compelling reason for selecting this method.

CONDUCTIVITY PROBES

The advantages of the conductivity switch include low cost, simple design, and elimination of moving parts in contact with the process material. It can also be used to detect the level of moist bulk solids.

It has several disadvantages. In chemical processing, the possibility of sparking when the liquid level is close to the probe often cannot be tolerated. However, some solid-state units are rated for intrinsic operations. Most switches are limited in application to conductive (below 10^8 ohm/cm resistivity) and noncoating processes.

In many processes, electrolytic corrosion at the electrode can be a problem. Electrolysis can be reduced, but not eliminated, by using ac currents.

DIAPHRAGM LEVEL DETECTORS

Diaphragm detectors operate on the principle of detecting the pressure exerted by the process material against the diaphragm. There are diaphragm switches for liquid and solid services and diaphragm devices for continuous liquid level detection.

The diaphragm switches for solids include devices with mercury contacts which can be used with materials having a bulk density of more than 30 lb/ft (0.5 g/m). Units with microswitches are used for lower-density services.

Some sensitive diaphragm switches can be actuated by 6 oz. (170 g) of force. The differential of a single diaphragm can be as high as 8 in. (203 mm), which means that the switch will close when the solids rise to the top of the diaphragm and open when they drop 8 in.

The lower the solid density, the larger the diaphragm area needed. Units are available with 4 to 10 in. (102 to 354 mm) diameter diaphragms. There are two basic ways of installing these detectors: (1) they can be suspended on a support pipe; (2) they can be mounted on the inside wall of the bin. The mounting location should always be selected to guarantee the free flow of solids to and from the diaphragm area.

Diaphragm switches can be used to detect liquid level by sensing the pressure of a captive air column in a riser pipe beneath the diaphragm. An 8-in. (203-mm) head of liquid above the inlet of the riser pipe will compress the air sufficiently for switch actuation, and many units can handle a maximum of 60 ft (18 m) of liquid. The diaphragm is in contact with the captive air but not with the process. These units are limited in use to atmospheric tanks, and should be considered only for secondary applications where low cost is desired and accuracy is not critical.

Continuous diaphragm level detectors for liquid service are used to sense the pressure of the liquid. These installations are limited to atmospheric tanks, and used in applications where low cost is important compared to the quality or accuracy of the measurement.

The diaphragm unit is similar to a riser pipe diaphragm switch except that the diaphragm isolates the captive air from the process fluid. The unit consists of an air-filled diaphragm box connected to a pressure detector via capillary tubing. As the level rises above the diaphragm, the liquid head pressure compresses the captive air inside. The air pressure in the capillary tubing is sensed by a pressure element and interpreted as a level indication.

DIFFERENTIAL PRESSURE LEVEL DETECTORS

Liquid level may also be measured by a differential pressure (d/p) instrument. For vessels operated at atmospheric pressure, one side of the instrument is connected to the bottom of the vessel (level pressure) and the other side is vented to the atmosphere (reference pressure).

For pressurized vessels, the pressure side is connected to the vapor space in the vessel. This will give an accurate measurement of the liquid level provided that the density of the process does not change.

The differential pressure can be detected by sensing the two pressures separately and obtaining the difference. Generally, it is more desirable to use a single-pressure differential sensor so that the static pressure levels are intrinsically balanced. For example, consider a 100-in. (2.5-m) water column measurement where the expected accuracy must be $\pm\frac{1}{2}$ in. (± 12.7 mm). If this measurement were made at a static pressure level of 1000 psig (6.9 MPa) using two independent sensors it would be very difficult to approach this accuracy because of differential errors between the devices.

The differental pressure device can take the form of a pneumatic transmitter. Here, a pair of diaphragms is welded to opposite sides of the pressure cell and the space between them is filled with liquid to damp out process noise, which is a common problem in flow applications, but not usually a consideration in level installations. The differential pressure to be detected is applied to the two sides of this diaphragm capsule. The resulting force due to a change in differential pressure causes a force bar to change its position and this causes a change in the pneumatic output signal. The change in air pressure is detected by a feedback bellows and the bar force is opposed by an equal force developed in the feedback bellows. As a result of this action, the output signal is maintained proportional to the differential pressure.

Electronic d/p transmitters use a force bar and fulcrum assembly similar to this arrangement; however, the motion of the force bar is detected using an electrical element such as a variable-reluctance or strain gauge transducer. The output signal may be fed back through a coil to rebalance the force bar in units that use a force-balance design.

Transmitters that do not use a force bar detect the position of the sensing diaphragm directly. Some devices use a variable capacitance. An increasing pressure on the high-pressure diaphragm causes oil to flow into the inner chamber, forcing the sensing diaphragm to one side. The capacitance between the high-pressure capacitance plate and the sensing diaphragm increases and the capacitance between the sensing diaphragm and the low-pressure plate decreases. This change is detected and used to produce a current proportional to the differential pressure.

These units do not use electromechanical feedback, but they are based on a motion balance design. Like the pneumatic d/p transmitter, the low side can be vented to atmosphere or it can be connected to the vapor space in the vessel for pressure compensation. Typical ranges are from 5 in. (127 mm) of water to 1000 psi (6.9 MPa) with temperature ratings up to 250°F (121°C).

IMPEDANCE PROBES

As we discussed earlier, capacitance probes may not perform satisfactorily if the process builds up on the probe to any significant degree. The impedance probe, which is a modification of the capacitance probe, is designed to overcome this problem. As material builds up on the probe, a low-resistance path to ground, the vessel wall, is formed. Since the switch detector circuitry cannot discriminate between the current flow in the resistive path and the current flow in the capacitive path, the switch will operate once the probe is coated.

One method that is used to overcome this problem is to use a bucking voltage in the resistive path to ground, thereby eliminating the resistive path current flow. A second probe, which is insulated from both ground and from the primary probe, is driven in phase and at the same voltage as the primary probe. The second probe breaks the resistive path to ground when the probe is coated; the switch may be set to trip when liquid rises to cover the probe, thus completing the capacitive path.

For the case of nonconducting processes, material buildup will not cause the formation of a resistive path to ground but will cause an unwanted capacitive path, causing a high-level trip in the absence of the process material. By using the secondary probe technique, the coupling effect of the buildup can in many cases be balanced out so that the effects are minimized.

The continuous level measurement of sticky conducting materials introduces an additional complication. The resistive path to ground at the vessel connection can be broken in the same manner as discussed above. But since a portion of the probe is always covered by the process, a second resistive path to ground is formed through the liquid to the vessel wall. In order to make a measurement under these conditions, the electronic circuitry in the transmitter must be able to discriminate between the resistive current flow and the capacitive current flow. Since these two are out of phase, it is possible to make this distinction in most cases.

If a continuous level measurement is to be made in a sticky, nonconducting material, there may be unwanted capacitive coupling near the surface of the liquid. There is no easy way to correct for this, so some inaccuracies in the measurement are usually tolerated.

The impedance probe, which overcomes some of the weaknesses of the capacitance probe, can be used in a wider range of level detection applications. The proper operation of the impedance probe and its accuracy can still be affected by coating thicknesses and changes in process dielectric or conductivity.

LEVEL MEASUREMENT USING DISPLACEMENT TECHNIQUES

Archimedes' principle implies that a body wholly or partially immersed in a fluid is buoyed up by a force equal to the weight of the fluid displaced. By detecting the apparent weight of an immersed object, a level instrument can be made.

A simple application of this principle uses an object that is heavier than the process liquid and is suspended from a spring scale. When the liquid level is below the object, the scale shows the full weight. As the level rises, the apparent

160 TRANSDUCERS FOR AUTOMATION

weight of the object decreases. The spring scale can be calibrated in percent, or in level units. This type of device is limited to use in open tanks.

In most industrial situations, a basic problem is to seal the process from the spring scale or other force detecting mechanism. The seal must be friction free and usable over a wide range of pressures, temperatures, and corrosion conditions. The variations in the design of this seal tend to distinguish the types of displacement detectors available. These include:

1. The magnetically coupled switch.
2. The torque tube.
3. The diaphragm and force bar.
4. The spring balance.
5. The flexible disk.
6. The flexible shaft.

Each of these devices uses Archimede's principles, but they differ in the type of seals employed. All of these can be used to detect the liquid-vapor interface, liquid-liquid interface, and, if the level is constant, density changes.

The flexible disk unit is available as a pneumatic transmitter, and the flexible shaft unit is available as a high-gain pneumatic controller. The other designs are available with local pneumatic controllers and pneumatic or electronic transmitters.

A float level switch stays on the surface while a displacer switch is partially or totally immersed. As shown in Figure 7-4, the displacer is usually mounted on a flexible cable which is attached to a support spring. When the tank is empty, the spring is loaded with the full weight of the float. The displacer switch has several advantages over the float switch:

1. Level settings or spans are easily adjusted by moving the displacer to a new elevation on the cable.

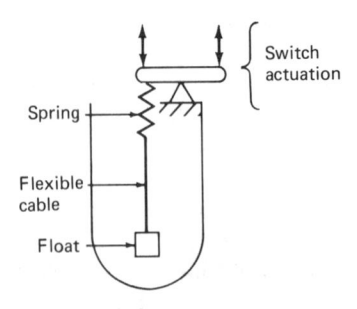

Figure 7-4. Displacer switch.

2. The maximum differential between high and low settings can be as great as 50 ft. (15 m).
3. Fluid density usually has no effect on the displacer diameter and units are interchangeable between systems of varying density by changing the support spring.
4. Surface turbulence is not as likely to cause switch chatter since the cable is in tension.
5. The displacer switch is less likely to cause spurious noise in vibrating service since the cable is always in tension. This is important for automatic shutdown.
6. The displacer switch is very flexible and can be used for multiple control functions.

The major disadvantage of the displacer switch is that the spring is exposed to the process. This limits the switch to applications that are clean, nonfreezing, and noncorrosive.

The torque tube displacer switch uses a torsional spring to support the float. The torque arm connects the float to the torque tube and absorbs lateral forces. Friction is minimized by use of a knife edge bearing support. A limit stop prevents accidental overstressing of the torque tube by limiting the downward motion of the torque arm. The angular displacement of the torque tube and torque arm are the same at the knife edge and at the tube. At the flange end, the tube does not rotate at all because it is solid, but the torque rod is free to rotate the same amount as it did at the knife edge. The angular displacement, which is usually 5° or 6°, is linearly proportional to the apparent weight of the float. The torque tube material normally is Inconel but it is also available in stainless steel, Hastelloy, Monel, nickel, or Durimet.

The torque tube is designed to twist a fixed amount for each increment of buoyancy change. Thus, in selecting the float diameter, the torque tube characteristics, the density of the process fluid, and the level span must be considered.

Torque tube level devices may be mounted internally or externally to the vessel. Internal devices are used in applications where the tank can be drained when the level detector requires maintenance. If the unit is to be internally mounted, it can be installed inside a stilling well which may be fabricated from pipe. The pipe should contain a number of vertical slots among its length and should have a stop bar welded across the bottom to prevent the float from sinking in the vessel if it becomes disconnected from the torque tube.

In installations where the vessel cannot be depressurized and drained to perform maintenance, the unit should be installed in an external chamber mounted outside of the tank, isolated from the process by isolating valves. Since the process is exposed to ambient temperature when an external chamber is used, it may be necessary to heat trace and insulate the installation for freeze protection.

The spring balance instrument is similar to the torque tube unit with two exceptions: (1) the torsional spring is replaced by a conventional spring; (2) the process is isolated from the instrument by means of a magnetic coupling. The float is suspended in the liquid by the spring. As the level in the vessel rises or falls, the effective weight of the float changes, causing the spring to extend and contract. A magnetic attracting ball attached to the displacer rod rises and falls in response to the displacer movement. The movement is about 1 in. (25 mm) full range. The ball is centered within the enclosing tube and its movement is relatively friction free.

Float level switches and indicators employ a float which follows the liquid level or the interface level between liquids of different specific gravities. The smaller-diameter floats are used in higher-density materials and the larger floats are used for liquid-liquid interface detection or lower-density materials. Float-operated switches and indicators fall into one of three categories:

1. Direct connected for atmospheric tanks.
2. Sealed units for pressurized tanks.
3. Tilt switches for liquids and solids.

Atmospheric tanks can use a unit as shown in Figure 7-5 which is basically a tape gauge. A tape is connected to a float on one end and to a counterweight on the other to keep the tape under constant tension. The float motion causes the counterweight to move up and down, indicating the tank level. The device is common in water storage tanks, although it can be used in any process that can be left open to the atmosphere. The float and tape materials must be selected to meet corrosion requirements. The range is a function of tape length and can exceed 100 ft (30 m).

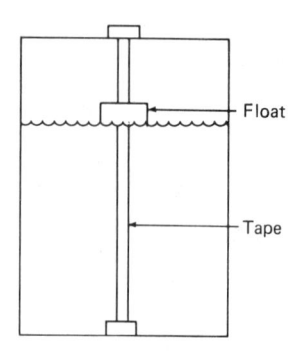

Figure 7-5. Inductively coupled float and tape level detector.

LEVEL MEASUREMENTS 163

Pressurized tanks require a seal between the process and the indicator or switch. The float motion can be transferred to the switch or indicator mechanism by magnetic coupling or other techniques.

In some devices, the vertical position of a rotameter float is used as an indication of flow rate. This device is modified for level indication by installing a solid shaft on the float in the process fluid. The shaft moves the detector in the rotameter housing with level change.

The float can be placed directly in the tank or can be mounted in an external chamber with isolating valves. The range of this device is about 15 in. (381 mm), which limits its applications. Materials are either steel or stainless steel, with metallic tubes used with magnetic coupling to drive the scale indicator. This design can be used for operating pressures to 1200 psig (8 MPa) and temperatures to 800°F (427°C).

The float and guide tube design shown in Figure 7-5 may use a series of magnetically coupled float switches that are employed for continuous level indication. In the most basic configuration, a reed switch is placed inside a sealed and nonmagnetic guide tube and the rising or falling liquid level is allowed to actuate the switch.

The float contains an annular magnet and is guided by the tube. A mechanical stop is placed on the guide to prevent the float from rising above the elevation of the switch. The stop allows the switch to stay closed whenever the level is at or above the switch elevation and it opens only when the process level falls. Several mechanically stopped floats can be placed along the tube to allow multiple switching.

The float and guide tube design can be employed for continuous level indication by using a large number of closely spaced switches inside the guide tube and detecting which one is being closed by the magnet in the ball float. A simplified scheme for this is shown in Figure 7-6. A voltage divider network is

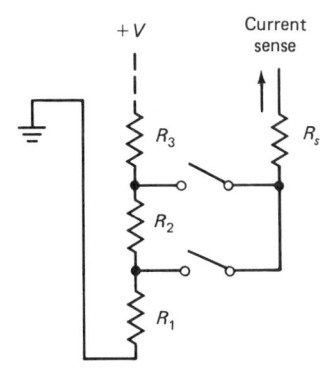

Figure 7-6. Voltage switching.

made by connecting resistors $R_1, R_2, \cdots, R_n$ in series across the power supply. The system can be scaled to read level in percent or in engineering units. Resistor R_s is made large so that the output current flow is small in comparison to that in the divider.

The switches can be placed as close as $\frac{1}{4}$ in. (6 mm). The replacement of a faulty switch can be a time-consuming task and for this reason the indicator is normally equipped with a redundant switch at each point. The maximum length for this design is 10 ft (3 m) if the assembly is used with a metal guide tube. It is also possible to use a flexible, plastic-jacketed assembly that is available in longer coils. These assemblies are field installed into a stainless-steel or other nonmagnetic guide tube. Another option available is solar-powered circuitry, which reduces the installation cost by eliminating the need for an electric power line.

Tilt switch designs are available for liquids or solids. The liquid design uses a mercury switch enclosed in a plastic case which is suspended from a cable at the desired level. When the liquid reaches the case, the case tilts, causing the switch to close or open. This device is used in sumps and ponds and is limited to atmospheric pressure applications. If it is installed outdoors above grade, it should be equipped with a wind screen.

Another tilt switch design is used primarily to detect the presence or absence of solids. Suppose it is desired to monitor the material on a conveyor belt. As long as there is material on the conveyor belt, this switch will be tilted up. If the material feed to the belt is lost, the switch rotates to a vertical position and the switch contacts change state.

Float-operated devices are not normally used for industrial pressure vessel applications. The reason is that for pressure vessel service the wall of the float must be thick enough to prevent it from collapsing. In many cases this results in an overly large float. Another problem is that immersed magnets will attract pipe scale and other ferrous metal particles in the process material that can interfere with proper switch operation. Many of the float-operated designs have close tolerance moving parts exposed to the process and cannot be used in dirty or plugging services.

The ball float designs tend to be inexpensive and reliable, which results in their selective use in industrial processing and wide use in an array of applications outside of this area.

LEVEL GAUGES

Level gauges provide a reliable and inexpensive visual indication of the level in tanks and vessels. The first designs were the tubular glass type. Flat glass, magnetically coupled, remote reading, and bull's-eye designs are also available. These designs are safer and can be used in more applications.

Flat glass gauge sections are available in 4 to 20 in. (101 to 508 mm) lengths and the overall gauge assembly can have as many as five sections. The transparent gauge has glass sections on opposite sides of the chamber and the reflex gauge has a single glass with prisms cut in the glass on the process side. Light striking the glass in the vapor phase is refracted, making the vapor space appear silver-white. Light striking the glass covered with liquid is refracted into the liquid, making that portion black.

The reflex gauge is used on clean, clear, noncorrosive liquids. Transparent gauges are used on interface services and when the process is dirty or viscous. Transparent gauges are also used when internal mica or plastic shields are required to prevent chemical attack or discoloration of the glass. This includes some caustics, hydrofluoric acid, and steam. Reflex gauge glass cannot be shielded because of the prisms that are cut in the glass. Some process materials may also act as a solvent and dissolve the coating on the reflex gauge, reducing the effectiveness of the prism.

OPTICAL LEVEL SWITCHES

There are several other devices that use light to detect the presence of a liquid. In one design a beam of light is aimed at the liquid and reflected back to a light-sensitive transistor (Fig. 7-7a). By adjusting the transistor sensitivity, the unit can be calibrated for point level detection from $\frac{1}{4}$ in. (6.3 mm) to 12 ft (3.6 m). The sensor can also consist of several light-sensitive detectors. The operating temperature range is -60 to $160°F$ (-51 to $71°C$).

The noncontacting design is suitable for use in corrosive, sticking, or coating processes; however, it is adversely affected by changes in the reflectivity of the process.

A light source and detector can be used to continuously monitor suspended solids in a liquid. Units based on this principle are also available for detecting sludge level. These units must only detect a boundary, so they are considerably simpler than continuous detectors.

In some units, the sensor and electronics are portable, and the sensor is furnished on a cable which is lowered into the vessel until the boundary is detected.

In another design (Fig. 7-7b) a light beam is directed along a cylindrical translucent rod that has a 45° bevel at the base. When no liquid is present at the tip, the beam is reflected across the rod and up to a light-sensitive transistor. As the level rises to cover the tip of the probe, the index of refraction increases and the amount of light received by the transistor is reduced, causing it to switch. The unit is sensitive to a change in level of $\frac{1}{16}$ in. (1.6 mm), has a pressure rating of 100 psig (0.69 MPa), and has a temperature range of 15 to 250°F (-90 to $121°C$).

166 TRANSDUCERS FOR AUTOMATION

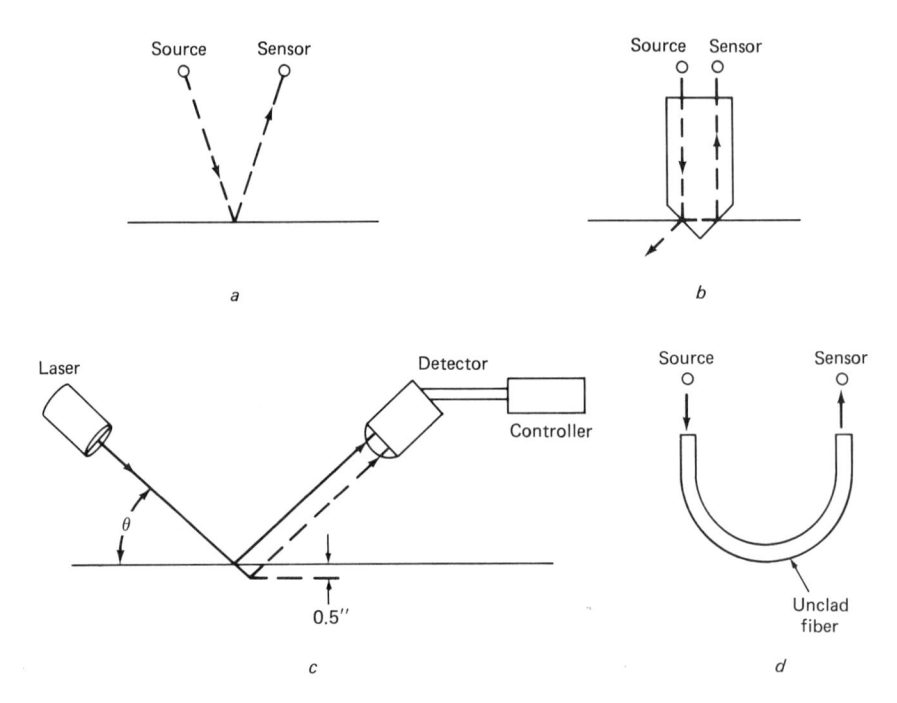

Figure 7-7. Optical level sensors. *a*. Noncontacting. *b*. Contacting. *c*. Laser. *d*. Fiber optic.

The switch cannot be used in caking or coating liquids since the presence of liquid on the probe after the liquid level has dropped can cause the switch to continue to indicate a high level.

A laser beam and detector may also be used for continuous level measurement. As shown in Figure 7-7c, a laser source is mounted on one side of the vessel at an angle θ between 15° and 60° from horizontal.

The detector is mounted on the other side of the vessels and at the same angle as the source. As the surface level changes, the beam is displaced as shown. This displacement can be measured and converted to level.

In order for this unit to work properly, the surface of the process must be clean and reflective. The span is limited to approximately $\frac{1}{2}$ in. (12.5 mm). This noncontacting technique can be used to monitor the thickness of such materials as molten glass as it is formed into sheets.

A fiber optic system can also be used for liquid level detection (Fig. 7-7d). A light beam is sent through the fiber and when there is no liquid on the fiber, the return beam will have the same intensity as the source beam. As the liquid covers the fiber, the index of refraction increases, reducing the strength of the return beam.

Optical level devices are not common in continuous chemical processing applications because of pressure and temperature limitations and because they do not operate well in fouling and caking services. They are useful for level detection in less severe environments.

RADIATION LEVEL SENSORS

Atoms with the same chemical behavior but with a different number of neutrons are known as isotopes. Many elements have one or more stable isotopes. For example, the stable isotopes of oxygen are

Isotope	Protons	Neutrons	Abundance (%)
O 16	8	8	99.76
O 17	8	9	0.04
O 18	8	10	0.20

Most elements also have unstable (radioactive) isotopes. Oxygen's radioactive isotopes are O 15 and O 19, which have 7 and 11 neutrons. The unstable isotopes disintegrate to form elements or stable isotopes.

Most of the elements that are heavier than lead are unstable and disintegrate to form lighter elements. Radioactive disintegration is accompanied by the emission of three different kinds of rays: (1) alpha radiation, which consists of positively charged particles having two neutrons and two protons; (2) beta radiation, which consists of electrons; and (3) gamma radiation, which consists of electromagnetic waves similar to x rays.

The relative penetrating powers of the three types of radiation are approximately 1, 100, and 10,000. The penetrating power of alpha rays is less than 8 in. (203 mm) of air. Alpha and beta rays have an electrical charge and can be deflected by an electric or magnetic field.

Gamma rays have a much higher penetrating power and cannot be deflected, so gamma radiation sources are normally used for level detection. The most common gamma sources are the radioactive isotopes of Co 60 (cobalt) and Cs 137 (cesium).

Co 60 is produced by bombarding the stable isotope Co 59 with neutrons. When Co 60 decays, it emits beta and gamma radiation to form the stable element Ni 60 (nickel). As Cs 137 decays it emits beta and gamma radiation to form the stable element Ba 137 (barium). Cs 137 is a fission product of uranium and is obtained from spent nuclear power fuel rods.

The source loses strength as it decays with the rate of decay expressed as the half-life or the period of time during which the source loses half of its strength. Co 60 has a half-life of 5.3 years and decays approximately 12.3% per year. Cs 137 has a half-life of 30 years and decays at a rate of 2.3% per year.

Cs 137 is normally selected for level measurements. Co 60 is used for applications where high penetration ability is required, as on vessels with thick walls. The main reason for this is its much longer half-life. In some installations, the isotope Ra 226 (radium) is used. This material has a half-life of 1602 years.

For point or continuous level measurement, the source decay does not affect accuracy, but the initial source must be large enough to allow the installation a useful life.

The unit used to describe the radioactive activity of any material is the curie (c). A gram of Ra 226 has 3.7×10^{10} disintegrations per second. This rate of activity is defined as one curie if it is produced by Ra 226 or any other source.

For most level detection applications source strengths of 100 milicuries (mc) or less are used. The unit of radiation is the roentgen (r), which is defined as the quantity of radiation that will produce an ionization equal to one electrostatic unit of charge in one cubic centimeter of dry air under standard conditions. A one-curie source will produce a dose of one roentgen at an object placed one meter away from the source for one hour. The dose rate is measured in roentgens/hour (r/h), which is based on the number of photons reaching the object at a defined distance.

Radiation is attenuated when it penetrates liquids or solids, and the rate of attenuation is a function of the density of the material in what is sometimes called the half-value layer. A 1-in. (25-mm) steel plate will reduce the radiation from a Cs-137 source by half. An additional 1-in. plate will cause another 50% reduction, so the overall reduction caused by a 2-in. (50-mm) plate is 25%.

The amount of radioactive material required to produce one curie of activity depends on the material. One curie is generated by one gram of Ra 226, 0.88 mg of Co 60, or 0.115 mg of Cs 137. The radiation field intensity in air can be found from

$$D = \frac{1000 \, KS}{d^2},$$

where

D = intensity (mr/h),
S = value of source (mc),
d = distance to source (in.),
K = constant: 1.3 for Ra 226,
    0.6 for Cs 137,
    2.0 for Co 60.

In most applications, the radiation must penetrate materials other than air, and we would like to determine the radiation field intensity after the gamma rays have passed through the vessel walls and process material. The previous

LEVEL MEASUREMENTS 169

equation and the appropriate transmission factors can be used for this purpose. Figure 7-8a shows a typical installation; θ is usually a maximum of 45°.

We would like to determine the radiation intensity at the receiver and what levels of operator exposure might be expected. Assume that the minimum radiation field intensity at the detector is 2.0 mr/h when the 3-ft-diameter vessel is empty and that the field should be reduced by at least 50° with the vessel full. The liquid in the vessel has a specific gravity of 1.0. Assume the use of a 100-mc source of Cs 137. Without the vessel the field intensity at the detector becomes

$$D = 1000 \frac{KS}{d^2} = \frac{0.6 \times 100}{84^2} = 8.5 \frac{\text{mr}}{\text{h}}.$$

With the empty vessel in place the attenuation through two $\frac{1}{2}$-in. steel walls will be $0.70 \times 0.70 = 0.49$ using Table 7.1 and the resulting field intensity at the

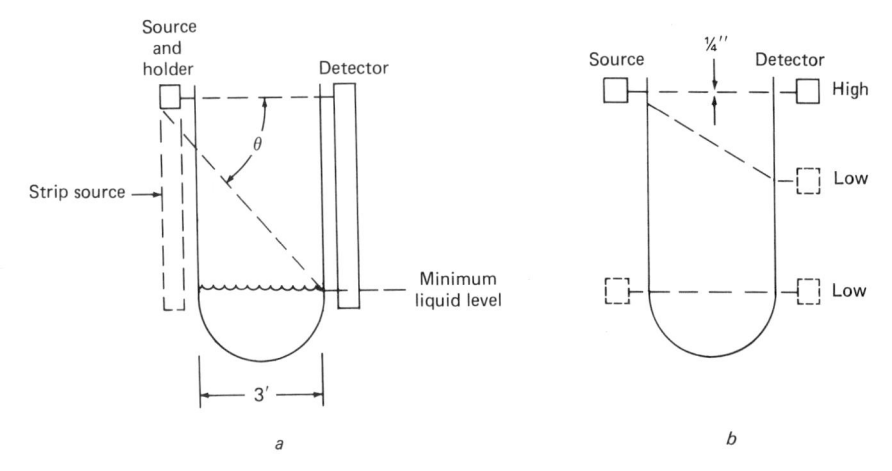

Figure 7-8. Radiation level sensing. *a*. Continuous level detection. *b*. Level switch installation.

TABLE 7.1. Transmission (%) of Cs 137 through steel and H$_2$O.

Material thickness (in.)	Steel	H$_2$O
$\frac{1}{2}$	70	
1	50	
2	25	
3	8	90
8		50
12		30

detector will be 8.5 × 0.49 = 4.2 mr/h. When the tank is full, the radiation will have to penetrate 36 in. (0.9 m) of material having a specific gravity of 1.

The total attenuation will be the product of the attenuation in three 12-in. (0.3-m) thicknesses. Each 12 in. (0.3 m) of material thickness has a transmission of 30%, so the field intensity at the detector with the tank full is

$$4.2 \times 0.30 \times 0.30 \times 0.30 = 0.113 \text{ mr/h.}$$

A gamma source radiates electromagnetic energy in all directions; short-term exposure to high-intensity gamma radiation or long-term cumulative exposure to lower-intensity radiation is hazardous. The degree of hazard, particularly to long-term low-intensity exposure, is somewhat subjective, but if an error is made it should be on the conservative or safe side.

The radiation sources are formed into ceramic pellets which are then placed in a double-walled stainless-steel capsule. The capsule is contained in a source holder which lets a radiation beam in through a narrow window since it is blocked by lead shielding in the other directions. A shutter allows the window to be closed when the source is being shipped or when it is out of service. The shielding must be thick enough to reduce the field intensity 1 ft (305 mm) from the source to 5 mr/h or less.

In the United States, rules governing the safe exposure rates to radioactive materials have been established by the Nuclear Regulatory Commission (NRC). These rules are incorporated in the Occupational Health and Safety Act (OSHA). Exposure to external radiation is referred to in "rem" units (roentgen + equivalent + man). A person receives a dose of one rem when exposed to one roentgen of radiation in any time period. A person should not receive more than 250 rems over an entire lifetime.

If it is not possible to isolate the area of high field intensity, additional lead plate shielding can be permanently installed around the source holder and/or behind the detector.

In most applications, the source and detector can be arranged or shielded so that the operator receives less than 10% of the safe levels, assuming a 40-hour per week occupancy. If a 1-in. (25-mm) thick lead shield is installed behind the source, the operator exposure would be approximately 75 mrem per year, assuming a 40-hour per week exposure.

Among the gamma radiation detectors available, the two commonly used for level detection are the Geiger-Mueller (G-M) tube and the gas ionization chamber. The G-M tube has a wire element anode in the center of a cylindrical cathode. The tube is filled with inert gas and sealed. A voltage of 250 to 300 V is impressed across the anode and the cathode. The incident gamma radiation ionizes the inert gas and produces a current between the anode and cathode. The frequency of the ionization current is related to the intensity of the gamma radiation and this can be determined by counting the pulses produced over a time interval.

The ionization chamber is also filled with inert gas and sealed, but instead of applying a high voltage, a much smaller voltage of 6 V is applied across the chamber. When the chamber is exposed to gamma radiation, ionization occurs and a continuous current in the microampere range flows. This current is proportional to the radiation intensity.

The G-M tube is most always used for level switch point detection. The switch detector is arranged so that it observes the full field intensity for the tank empty condition or observes little or no field when the tank is full.

Both types of detectors are used for continuous level detection. Ionization chambers are available in continuous lengths of up to 10 ft (3 m). G-M tubes are available in 6-in (152-mm) or 12-in. (305-mm) lengths, and can be stacked to form a continuous detector.

The stacked G-M tubes are less expensive than an equivalent length of ionization chamber, but the G-M tubes are more sensitive to drift and they can deteriorate with time. If a G-M tube is exposed above the liquid level fails, it will appear to the receiver that the failed section is covered by the liquid. In a four-section assembly this would cause a reading 25% high.

A set of typical level switch installations is shown in Figure 7-8b. In the most common installation the source and detector are at the same elevation and the detector is horizontally mounted. The differential between on-off action is about $\frac{1}{4}$ in. (6.3 mm), which means that a $\frac{1}{4}$-in. rise in liquid level is enough to block the source beam and change the state of the switch. For a wider differential, the detector can be mounted at an angle to the horizontal. For the maximum differential, the sensor is mounted vertically, producing a differential of about 6 in. (152 mm). For even wider differentials, two detectors can be used with a single source. The maximum differential between high and low level settings is then equal to the tank diameter. Differentials greater than the tank diameter can be achieved with two separate sets of sources and detectors.

In high-level switching applications, the G-M tube and switch are normally above the liquid level where they are exposed to the full field intensity. Pulses from the G-M tube are sent to a trigger circuit that continuously resets a time-out much like the time-out used in a computer watchdog circuit. When a rising level blocks the radiation beam, the time-out is no longer reset and the switch changes state; for fail-safe operation, the switch would open. The switch circuitry is arranged so that the switch will open on failure of the G-M tube or failure of any of the switch components, so that the entire installation is fail-safe.

In low-level switching applications G-M tube or switch component failure would not be detected since exposure of the tube to the beam on falling level would not actuate the switch. For a fail-safe design for falling-level applications, a test circuit is required to test the switch when the level is high. One technique employs a small source in the detector to test the integrity of the tube and switch circuitry.

There are two techniques for continuous-level measurements with fixed sources and detectors. One method uses a strip source and detector, as shown in Figure 7-8a. The second, also shown in Figure 7-8a, uses a point source and a strip detector.

The first type, with the strip source, is more expensive, but it is more accurate and can be used for a wider array of vessel geometries than the point source detector. The strip source radiates a narrow, uniform beam in the direction of the detector. As the level rises, a small part of the detector is screened. The response is uniform and linear over the entire span except for small nonlinear and effects near the 0 and 100% points.

The strip technique is not as sensitive to variations in the specific gravity of the process material. If the specific gravity of the material varied from 0.40 to 0.70, the source could be sized for 0.40 specific gravity material. Material of higher gravity would cause a higher attenuation, but this would not affect the performance of the strip detector.

Radiation sensors can be used to detect, at a point or continuously, solids levels and liquid/liquid interfaces. The accuracy of the installation will depend on source size, detector sensitivity, material gravity, and vessel geometry.

Some applications may require the continuous monitoring of a liquid/liquid interface, or a liquid or solid level over a long vertical straight side. The strip source or multiple source and strip detectors needed would be expensive. But a point source and point detector can be motor-driven over a wide span to detect these levels. The motor drive can be designed to continuously hunt the level or to look for the level on operator demand, as required for an inventory.

The calibration of radiation level detectors is relatively simple. A 100% full calibration reading can be made by closing the source shutter to simulate a full vessel. The calibration of the actual level is done by running a portable G-M counter down the vessel wall between the vessel and the detector. The level is found when the counter reading drops off, since this is the point at which the liquid screens out the source.

Radiation level detection is useful for hard-to-handle, toxic, and corrosive materials since it does not require vessel wall penetration. Safety and licensing requirements cannot be minimized, but they are not a serious impediment in a carefully designed system. The characteristics of typical radiation level sensors are given in Table 7.2.

TABLE 7.2. Radiation level sensor characteristics.

Temperature range error	~20 to 160°F (−30 to 70°C) ±0.25 in. (±6.3 mm) switching ±1% continuous
Range	20 ft (6.1 m) stationary units 50 ft (15.2 m) motorized units

RESISTANCE TAPES

As shown in Figure 7-9, the resistance tape detector responds to level changes by changing the detecting circuit resistance. As the level rises in the tank, the hydraulic pressure of the material compresses the jacket, causing contact between the resistance element and the conducting base strip. The sensitivity of the tape is about 0.2 psi (1.4 kPa), which is the pressure required to cause contact of the tape section under the material surface. Above the surface, the resistance element is measured to detect the material level.

Since 0.2 psi (1.4 kPa) pressure is required to compress the jacket, the tape is never in contact all the way up to the surface of the process material. The uppermost electrical contact is some distance below the surface. This distance, which is called the actuation depth, is a function of the density of the process material. For water, the actuation depth is approximately 6 in. (152 mm); for lighter materials, it is greater. At calibration, this offset is compensated for by adjusting the zero setting.

In order to maintain accuracy, the pressure inside the tape jacket must equal the pressure in the tank. For atmospheric tank applications, this can be done by venting the tape to the atmosphere using a small desiccant dryer. For tanks

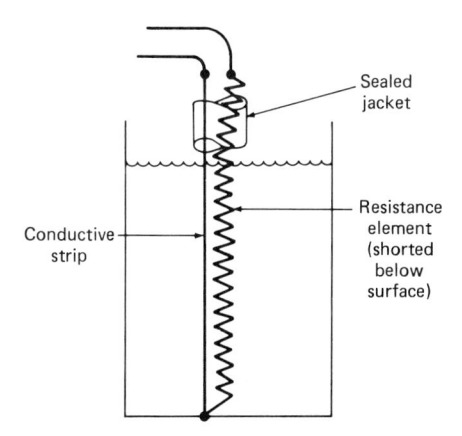

Figure 7-9. Resistance level detector.

under pressure or vacuum, the tape internal and external pressures must be equalized. This may be done by installing an equalizing line, or by connecting a pressure repeater between the tank and the vent connection on the tape.

In either case, care must be taken to keep moisture or other contaminants out of the tape jacket. As with many other level detectors, changes in the material's specific gravity can affect accuracy.

Resistance tape devices can be used for both liquid and solid services. The tape may be suspended through an access hole in the tank with a weight on the lower end, or it can be attached to the bottom of the tank. It may also be installed in a stilling well if the tank is to be agitated or if the tape requires mechanical protection.

A number of readout devices can be used for resistance tape elements. These range from battery-powered units to single or multichannel remote receivers.

Resistance tape units can be used to measure the level of solids, slurries, and liquids that are corrosive to other level detection devices. However, there are some limitations due to actuation depth, pressure equalization, and dryer maintenance problems. The characteristics of typical resistance tape level sensors are given in Table 7.3.

ROTATING-PADDLE SWITCHES

The rotating-paddle-type level switch is used to detect the presence or absence of solids in a silo. A small, geared, synchronous motor keeps the paddle in motion at very low speed. There is no torque on the paddle drive assembly when solids are absent. As the level rises to the paddle, the paddle is stopped and a torque is detected on the drive assembly.

One method used for torque detection is a modification of the displacer-type torque tube. When solids are not present and there is no torque on the tube, the entire drive assembly rotates at the gear speed. With solids present the paddle stops, a torque develops on the tube, and the torque rod is used to operate a switch.

In some units, the motor operates in a continuously stalled condition. In others, the motor is off and does not start until the torque detector returns to the zero position.

TABLE 7.3. Resistance tape level sensor characteristics.

Temperature range	5 to 160°F (−15 to 71°C)
Pressure	Atmospheric
Error	±1 in. (±25 mm)
Range	100 ft (30 m)

Larger paddle areas are required to generate sufficient torque in lower bulk density materials. Four blade paddles 6 in. (152 mm) wide by 2 in. (50 mm) high are used in materials with densities below 20 to 30 lb/ft^3 (0.32 to 0.48 g/cm^3). Vanes and wire blades are used for heavier materials.

Pressure ratings range from 7.5 psig (52 kPa) to 30 psig (207 kPa). The most common temperature rating is -30 to $200°F$ (-34 to $93°C$) although higher ratings are available.

Although the accuracy for these devices can be ± 1 in. (± 25 mm), the important factors for solids monitoring include (1) the location of the paddle, (2) the angle of the paddle, and (3) the possibility that the material will bridge the switch.

There are several methods of installing these switches. Units mounted in the side of a silo may be in the path of falling material so a protective baffle is installed over the unit to prevent spurious trips. Top-mounted designs generally employ a shaft guard. The characteristics of typical rotating-paddle sensors are given in Table 7.4.

TAPE LEVEL DEVICES

Tape level devices are often used with remote readout devices, making them suitable for inventory control in multiple tank installations. There are several devices to consider:

1. The inductively coupled float and fixed-tape design.
2. The wire-guided float detector.
3. The wire-guided capsule that is heavier than the liquid which uses a thermal sensor for level detection.
4. The surface sensor for liquids or solids.

The first three detectors are used for liquid level measurements and can have a resolution approaching ± 0.1 in. (± 2.5 mm). When used in liquids, the surface sensor can be as accurate; for solids its resolution approaches ± 1 in. (± 25 mm). For most solids applications, this resolution is adequate, since larger errors can be introduced in the correlation between level and volume by an uneven solids level and bridging.

TABLE 7.4. Rotating-paddle level sensor characteristics.

Temperature range	to 250°F (121°C)
Pressure	to 30 psig (0.2 MPa)
Error	± 1 in. (± 25 mm)

176 TRANSDUCERS FOR AUTOMATION

INDUCTIVELY COUPLED FLOAT AND TAPE DETECTORS

In this type of fixed-tape, float-actuated level measuring device, the tape is suspended into the tank and anchored to the bottom (Figure 7-10). The tape acts to guide a float, which contains an inductively coupled transducer.

The tape consists of a steel ribbon with a number of insulated conductors encapsulated in a Teflon jacket. In addition to providing mechanical strength, the steel tape is used to provide power to the transducer in the float through inductive coupling.

The conductors can be arranged on the tape in a coded pattern so that each 0.1-in. (2.5-mm) increment has a unique code. A receiver at the top of the tank detects which conductors are inductively coupled and this information is used to determine where the float is located and thus the elevation of the liquid level.

The float is Teflon coated and the tape-to-float clearance is usually $\frac{1}{4}$ in. (6.3 mm), to reduce float friction due to material buildup on the tape. The digital signal that is produced by the tape and float assembly has a resolution of 0.1 in. (2.5 mm) and can be used directly in a microcomputer control system.

The conductors on the tape are usually arranged to produce a gray code or cyclic binary output. In the gray code (Table 7.5) only one digit or bit in the word changes from one word to the next and on the tape only one conductor must change its position from one 0.1-in. (2.5-mm) increment to the next.

The number of conductors required increases with the span of the liquid level to be measured. The span covered is equal to 2^n, where n is the number of conductors. For four conductors, the span is 16 increments, or 1.6 in. (40 mm). If 14 conductors are used, the span is 16,384 increments, or 135 ft (41 m). A reference conductor and a return conductor are required. Figure 7-11 shows how four conductors might be arranged on the steel tape to produce the gray code digital word for a 16-increment measuring system. Each additional conductor

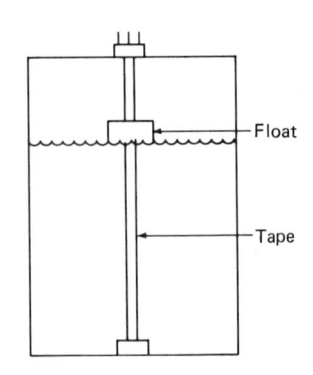

Figure 7-10. Tape level sensing.

TABLE 7.5. Gray code.

Increment	Output
0	0000
1	0001
2	0101
3	0100
4	0110
5	0111
6	1111
7	1110
8	1100

doubles the span; adding a fifth conductor would allow a measurement over 32 increments and a sixth conductor 64 increments. The characteristics of typical tape level sensors are given in Table 7.6.

WIRE-GUIDED THERMAL SENSORS

Liquid level can also be detected with thermal sensors. Since liquid conducts heat better than vapor, the liquid level can be sensed by two vertical thermal sensors. The lower one will always be cooler than the upper one.

The sensor must be heavier than the liquid, so the unit can be lowered to the

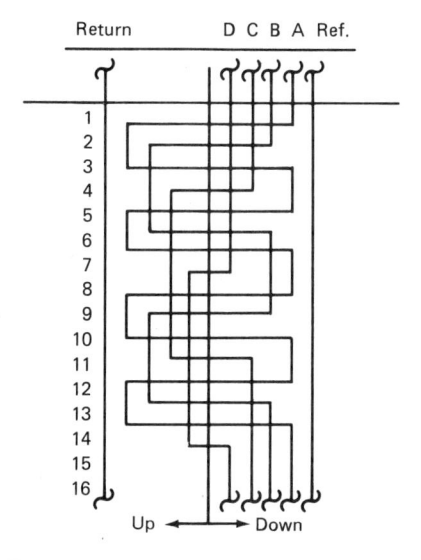

Figure 7-11. Gray code conductor layout.

TABLE 7.6. Tape level sensor characteristics.

Temperature range	Cryogenic service to 300°F (150°C)
Pressure	to 300 psig (2 MPa)
Error	±0.01 ft (±2.5 mm)
Range	100 ft (30 m)

tank floor. The cable tension must be sensed to signal when the level sensor reaches the bottom of the tank.

In a typical installation for a wire-guided thermal sensor system, the sensor, which is heavier than the liquid being measured, is suspended from an armored control cable and guided by a wire attached to the top and bottom of the vessel. A control unit detects the position of the sensor relative to the liquid level, and issues up or down commands to move the control cable until the lower sensor is in the liquid and the upper sensor is in the vapor. Generally, as the sensor is moving, each stepping command adds or subtracts a length unit from the previous reading so that the sensor position, and the level, is always known.

The sensor can also be used to provide a thermal profile of the tank material. This can be used for corrections of the level measurement and to detect temperature inversions in cryogenic service.

SURFACE SENSORS

This type of sensor was originally designed for solid level detection, but it can be used for liquid level detection with the proper sounding device. The sounder weight is suspended from a drum and the cable tension is continuously detected. A reversible motor rotates the drum until the sounder strikes the solid or liquid surface. As this occurs, the tension in the wire changes; this reverses the motor, returning the tension to its original value. Typically one counts down from a preset maximum reference value. As the sounder reaches the product surface, the level is recorded.

In dust-filled environments, a pneumatic cleaning assembly can be used to ensure reliable operation. There should not be any contact between the filling stream and the sounder. If the inlet nozzle is in the center of the bin, the surface of granular products will tend to take the shape of a cone. The sounder, then, should be located to obtain the average level.

There are other types of tape level detector designs available. In one design, the sensor is suspended on a cable and held above the liquid level. This distance is sensed by a proximity capacitance probe. A control unit monitors the capacitance between the sensor and the liquid level, positioning the sensor as the level changes. The sensor position, and the level, is a function of the amount of cable that is required.

Another design uses a displacer mounted on the end of a cable. The displacer is continuously repositioned so that it is always immersed the same amount. The level is a function of the amount of cable required. The displacement may be affected by changes in liquid density. Both of these designs should be installed in stilling wells.

Thermal level sensors can also be used in on-off services. The technique uses a resistive heater element that has a current flowing through it. A temperature switch is used to monitor the temperature of the probe. When the probe is covered, the heat generated by the resistive element is transferred to the water and the element remains close to the water temperature. If the level falls below the probe, the probe temperature will rise since the water vapor or air above the water has a lower thermal conductivity. The temperature switch senses this rise and indicates the level.

In another thermal level switch design, two temperature-sensitive probes are located in a horizontal plane such that the rising liquid level reaches them at the same time. One of the temperature elements is heated. When the probes are in vapor or air, the heated probe will be much warmer than the reference probe and its resistance will be higher. When the liquid covers both probes, their temperatures will be close to that of the liquid and their resistances will be almost the same. The sensitivity of this unit is high and it can be used for liquid/liquid interface detection or as a flow switch.

Another technique uses two resistive elements that are vertically displaced from each other. When a voltage is applied to both elements and they are both in vapor or in liquid, the heat transfer from the resistors to the media will be the same and their resistances will be the same. If the elements are connected in parallel, the current division can be compared to detect the level. If they are connected in series then the voltage division can be compared.

This design may be used as a continuous-level detector for a liquid/liquid or liquid/vapor interface in a tank. The system is adaptable to high-accuracy inventory monitoring for storage tanks.

The advantages of thermal detectors are (1) they contain no moving parts; (2) they are based on easily understood principles; (3) they can be very sensitive.

Thermal level detectors cannot be used on caking or plugging materials, nor can they be used when the additional heating will cause product degradation. Some units can only be used in nonhazardous services, since they require 120-V power.

VIBRATING-REED SWITCHES

The basic vibrating-reed level switch consists of a driver, paddle, and pickup. The driver coil normally imparts a 120-cycle-per-second vibration to the paddle, which is damped when the paddle is covered with process material. The pickup end contains a permanent magnet and a coil that generates a millivolt output signal when the paddle is vibrating. When the paddle is covered, the generated signal decreases. The unit will detect liquid/liquid, liquid/vapor, or solid/vapor interfaces since it is sensitive enough to detect relatively small changes in the density of the surrounding material. In many cases the switch will actuate before the paddle is fully covered by the process material. In water, switch actuation occurs when the water level is about 0.4 in. (10 mm) above the bottom of the paddle. In light powders (40 lbm/ft^3, the switch will actuate only when the level builds up to $\frac{1}{4}$ inch (6 mm) above the top of the paddle.

The vibrating-reed level switch can be used to sense both rising and falling levels, and it may be installed in tanks or pipelines. Its use for measuring density and viscosity has previously been discussed.

When the process material has a tendency to adhere to the paddle, the buildup can be removed by periodic purging, but these units should not be used where material buildup is likely. When used on wet powders, the paddle has a tendency to create a cavity in the ganular solids. This can result in a false level detection, since the vibration amplitude will soon be the same as if the paddle were in the vapor space.

Another design uses a self-sustaining resonant frequency probe. When the probe is in the vapor space, it will vibrate at its natural mechanical resonant frequency, which is about 100 cycles per second. One coil in the probe is used as a driver and a second coil is used to generate a feedback signal which is amplified and delivered to the drive coil. As the process fluid rises to cover the probe, it dampens the vibration and reduces the feedback signal. Without sufficient drive voltage the oscillation ceases.

The main advantage of this design is that material buildup on the probe has little or no effect, since the coating changes the natural frequency of the probe but does not completely stop the oscillation. Therefore, the sensor is suitable for slurry or similar services. This design is applicable for slurry services as well as low-density solid level sensing.

Vibrating-reed switches can detect the presence of materials with bulk densities as low as 1.0 lb/ft^3 (0.016 g/cm^3). The major disadvantage is that the swich setting cannot be easily changed.

ULTRASONIC LEVEL DETECTORS

Ultrasonic level sensors may be used for both continuous and point measurement. Point sensors are available for the measurement of gas/liquid, liquid/liquid, liquid/form, or solid/gas interfaces. These can be either damped-sensor or on-off transmitter types. Continuous level sensors may be used in or above the liquid. A 20-khz or higher oscillator is used as the ultrasonic signal generator.

Some units employ filters or other discriminatory circuits to prevent false readings that might be caused by random noise.

DAMPED-SENSOR LEVEL SWITCHES

These devices operate on the same principle as the vibrating-reed switch. When the sensor is in the vapor of the tank, it vibrates at its resonant frequency. It is damped out when it is in contact with the process material. Some units have a piezoelectric crystal in the vibrating tip.

Typical installations are shown in Figure 7-12. Note the top-entry installation where the vibrating face is in the vapor space and therefore undamped. Another installation does not penetrate the tank wall and is not in contact with the process media. Units may also be side-mounted, damped, or undamped. These units are normally limited to liquid service since the damping effect of many solids is not sufficient. The outside-mounted units can be used on any liquid, while the others are limited to clean, noncoating fluids.

ON-OFF TRANSMITTER LEVEL SWITCHES

These ultrasonic units contain transmitter and receiver sections. The transmitter provides the ultrasonic pulses and the receiver detects the pulses. The transmitter and receiver may be mounted on the same probe or they can be located on the opposite sides of the vessel.

Reflectors are sometimes used to narrow the beam angle when the distance between source and receiver is more than 10 ft (3 m). Some designs allow the measurement of both high and low levels with the same probe. Some probes

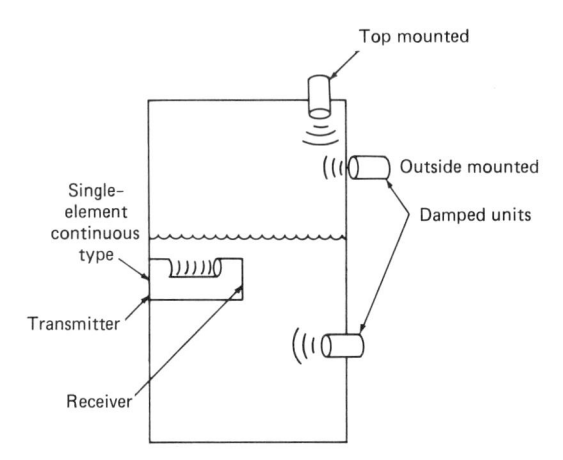

Figure 7-12. Ultrasonic level sensors.

are mounted at about 10° from the horizontal and used for the detection of liquid/liquid interfaces. When an interface is present in the probe cavity, the interface reflects the signal and prevents it from reaching the receiver.

These units can be used for clean liquids and none are suitable for slurry or coating service. The receiver and transmitter must generally be separated to detect the level of solids. These devices are all point sensors and they will not take into consideration the angle of repose during filling or emptying, nor will they detect ratholes, arches, or bridging.

CONTINUOUS LEVEL DETECTORS

These units depend on the echo effect of ultrasonic pulses. The time needed for the echo to return is converted to an indication of the level. The continuous ultrasonic level sensor measures the time required for an ultrasonic pulse to travel to the process surface and back. The source is an ultrasonic transducer or speaker and the receiver is a metal disk which acts as a microphone, being electrically and mechanically resonant. These sensors can be mounted either below or above the liquid level. Some designs use a two-element detector, where the transmitting and receiving transducers are packaged separately; other versions are packaged as a single unit.

The transducer is designed to generate a short burst of ultrasonic energy while the receiver is gated off. As the ultrasonic waves are released, the receiver is gated open to detect the echo.

For liquid level applications, the aiming angle should be within $\pm 2°$ of vertical. When measuring the level of solids, the angle of repose must be considered.

Some units employ a calibration bar for continuous calibration. These units can have high accuracy, provided that the density of the medium is uniform.

By using several sensors in the same bin, a visual profile can be obtained, showing the angle of repose and indicating whether the bin is being filled or discharged.

Ultrasonic level detectors can be used with success on troublesome solids since the frequency can be adjusted to suit the requirements of the process conditions. Other advantages include the absence of moving parts and the capability for continuous measurement without contacting the process medium. In some designs, penetration of the tank can be avoided. The measurement does not depend on detecting process properties and thus the reliability of the instrument is not affected by variations in composition, density, thermal or electrical conductivity, capacitance, or other properties.

The accuracy of the continuous unit can be improved to $\pm 0.1\%$ with the use of a temperature compensator. The ultrasonic level detectors are flexible and easy to install, and their accuracy is close to that of some accounting level units.

EXERCISES

1. Define what is meant by point level detection.
2. Discuss how solids can cause level measurement problems.
3. What are accounting-grade measurements?
4. What are the limitations of antenna probes?
5. Discuss the operation of bubble systems with the aid of a diagram. What is the major disadvantage of this type of level sensor?
6. Where is the capacitance type of level sensor most useful?
7. What conditions must be avoided when using conductivity level sensors?
8. Discuss the operation of the differential pressure level sensor. What are the installation requirements?
9. What are the advantages of the impedance type of level sensor?
10. Discuss how a displacement device could be provided with an electrical output. Give examples.
11. What are the safety requirements in using radiation level systems?
12. Discuss the operation of resistance tape detectors. What are the installation requirements?
13. What optical techniques can be employed to measure density?
14. Discuss how rotating-paddle switches operate. What are the requirements for accurate measurements?
15. What are the advantages of tape level devices?
16. Discuss the use of wire-guided thermal sensors. What are their advantages?
17. What applications are suitable for vibrating-reed switches?
18. Where are ultrasonic level detectors used?
19. Discuss the limitations of damped ultrasonic level sensors. What considerations are important to the application of the units?
20. What ultrasonic techniques can be used to obtain a visual profile of a solid material in a bin? Use a diagram to illustrate each method.

8
DISPLACEMENT AND PROXIMITY MEASUREMENT

APPLICATIONS

Sensors for the measurement of displacement and proximity are an important part of many automation applications. Resistive, capacitive, inductive, or optical methods can be used to measure the linear or angular displacement between the point or object being sensed and a reference or fixed point or object. Proximity sensors will measure linear or angular motion without any mechanical linkage.

The output from a displacement or proximity sensor may be an analog or digital equivalent of the absolute distance being sensed or it may be a function of the distance from a given starting point. Many of the common displacement and proximity techniques are used as primary sensors in other transducers such as many of the pressure types considered in Chapter 3.

RESISTANCE SENSORS

Variable resistors can be used as voltage or current dividers to provide displacement information. The wiper or movable arm of the resistor slides over the resistance element. Sensors are available for both linear and angular measurement including fractional and multiturn operation. The potentiometric displacement transducer is a low-cost device but it is not as accurate as other devices because of its mechanical design.

The resistance elements can be wire wound, carbon ribbon, or deposited conductive film and the excitation can be either ac or dc with no output amplifiers required. The output can conform to one of a variety of functional characteristics such as linear, sine, cosine, exponential, or logarithmic.

The disadvantages of this type of sensor are

1. Finite resolution for the wire-wound types.
2. Friction and limited life due to contact wear.
3. Increasing electrical noise due to wear.
4. Sensitivity to shock.

Typical devices may have the following characteristics:

Resolution—0.2%
Linearity—±1%
Repeatability—±0.4%
Hysteresis—0.5-1%
Temperature error—±0.8

Precision instruments are available with total errors in the range of ±0.5%.

CAPACITIVE SENSORS

Capacitive sensors are more often used for linear than for angular proximity measurement. In most devices either the dielectric or one of the capacitor plates is movable. The capacitive proximity sensor may use the measured object as one plate, while the sensor contains the other plate. The capacitance changes as a function of the area of the plates, the dielectric, or the distance between the plates.

Capacitive transducers are available with packaged signal conversion circuitry for dc output operation. Accuracy for small displacements may be nearly 0.25%, with accuracies of up to 0.05% available at a higher cost.

Capacitive devices are accurate, relatively small devices and have an excellent frequency response. The main drawbacks are their sensitivity to temperature and the need for additional electronics to produce a usable output. A typical capacitor transducer circuit is shown in Figure 8-1. An ac voltage is applied across the plates to detect changes. The capacitor can also be made part of an oscillator circuit in which it affects a change in output frequency.

Capacitive sensors have good linearity and good output resolution. The drawbacks are in the temperature and cable sensitivity, which requires the amplification circuitry to be located close to the transducer.

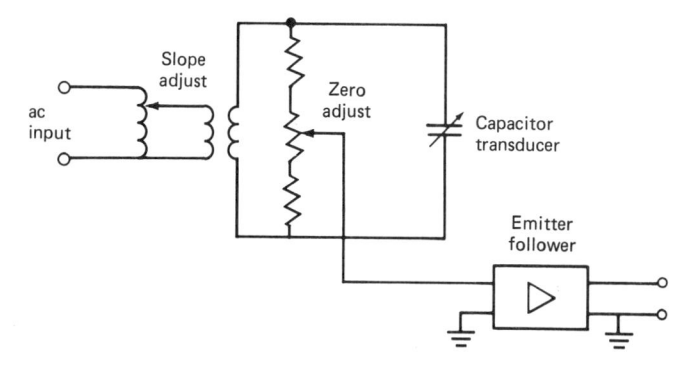

Figure 8-1. Capacitor transducer conditioning.

INDUCTIVE SENSORS

Inductive sensors may use a single coil in which a change in the self-inductance of the coil occurs. Multiple-coil designs use the change in magnetic coupling or reluctance between the coils. Single-coil displacement sensors use a movable core to change the self-inductance, while single-coil proximity sensors use the magnetic properties of the object itself to modify the self-inductance as shown in Fig. 8-2. The change in inductance is usually sensed with a bridge circuit or oscillator. Multiple-coil inductive sensors use the differential transformer technique and its variations.

The linear variable differential transformer (LVDT) uses three windings and a movable core to sense linear displacement. A typical LVDT configuration is shown in Figure 8-3.

In the LVDT secondary windings are wound to produce opposing voltages and connected in series. When the core is in a neutral or zero position, the voltages induced in the secondary windings are equal and opposite and the net output is a minimum. (See Fig. 8-4.)

As the displacement of the core increases, the magnetic coupling between the

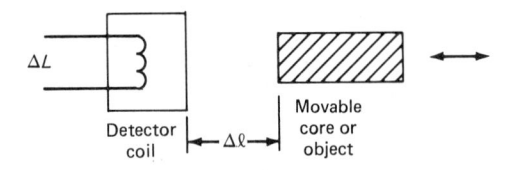

Figure 8-2. Inductive displacement and proximity sensors.

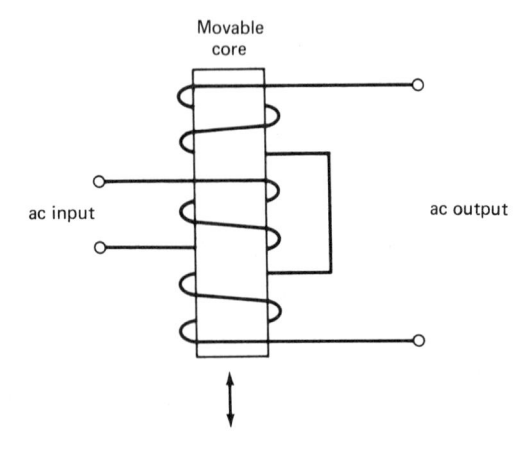

Figure 8-3. LVDT displacement transducer.

DISPLACEMENT AND PROXIMITY MEASUREMENT 187

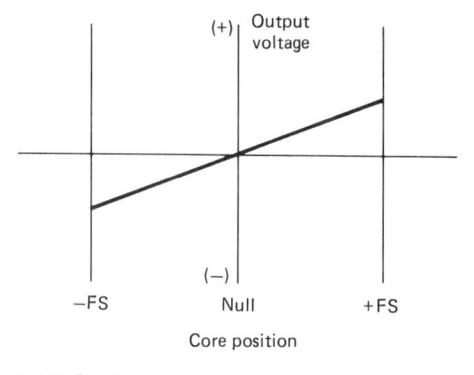

Figure 8-4. LVDT output.

primary coil and one of the secondary coils increases and the coupling between the primary coil and the other secondary coil decreases. The net voltage increases as the core is moved away from the center position and the phase angle increases or decreases depending on the direction.

A demodulator circuit is used to produce a dc output from the transformer windings. Input frequencies range from 60 Hz to 30 kHz. Some units have built-in conversion circuitry which allows dc input and output.

Differential transformers are also available in an angular arrangement in which the core rotates about an axis. Variations in the winding configurations are used in synchros, resolvers, and microsyns.

Inductance bridge sensors use two coils with a moving core to change the inductance of the coils, which form one half of an ac bridge. These sensors are available in both linear and angular configurations.

The temperature error for variable-reluctance transducers is typically 2% for a 100°F change. Transducers without the ac-dc conversion circuitry can operate up to 350°F, while transducers with conversion circuits are limited to −65 to 200°F. Typical linearity is ±0.5%. Repeatability and hysteresis can be less than 0.2%.

Variable-reluctance transducers offer good shock and vibration characteristics along with good dynamic response, but the ac conditioning circuitry required results in increased costs and reduced system reliability.

DIGITAL MEASUREMENT TECHNIQUES

Another type of position transducer which is important in automation application since it simplifies interfacing to displays and data acquisition equipment is the purely digital transducer. These devices have either a digital frequency or a digital-coded output which is a function of either displacement or proximity.

188 TRANSDUCERS FOR AUTOMATION

Those transducers with a frequency output use a frequency control that is a function of sensing movement. The transducers with a digital-coded output detect position and convert this deflection into a digitally coded word.

The digital output allows easy interfacing with other digital components and more accurate processing, resulting in greater system accuracy.

The digital output or series of pulses may be produced in displacement and proximity sensors using changes in electrical conduction, induction, or photoelectric conduction. Conducting encoders use brushes or wipers to detect the position of a coded disk or plate.

When a single sensing track is used, a series of pulses is produced as the disk or plate is moved. Direction sensing can be accomplished by adding another track which is offset to produce sequence logic. Counting circuits can be used to count the number of pulses and perform the conversion to angular or linear measurement. Multiple-track encoders produce a digital- or binary-coded output which is a function of the absolute angular or linear position.

Rotary encoders evolved from complex rotary switches, which produced multiple outputs or combinations of outputs as the switch position is rotated. A logical step was to adapt the switch outputs to fit a coded pattern which would indicate the position of the shaft.

The first shaft encoders used wipers or brushes and conductive disks. The disks were later plated with precious metal alloys for good conduction and in some cases the wipers were also plated with precious metal alloys to reduce arcing and noise. The use of these precious metal alloys was costly, and the conducting encoder, with its limited life, was used primarily in those applications where replacement was easy and cost was not critical.

Magnetic displacement sensors use gears of ferromagnetic material to produce pulses from a change in linear or angular position as shown in Figure 8-5. Direction sensing is accomplished by shaping the gear teeth in an asymmetrical pattern in order to modify the output waveform.

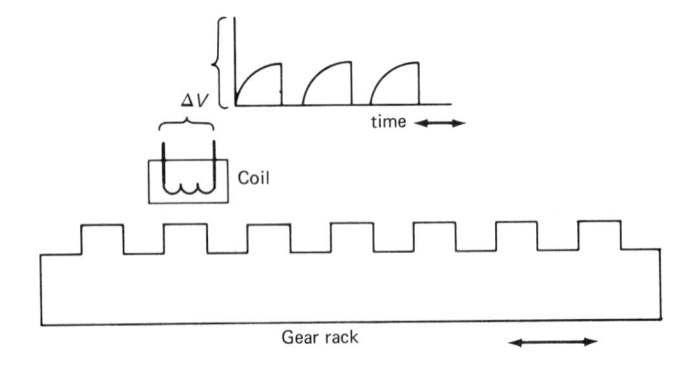

Figure 8-5. Magnetic displacement sensor.

Photoelectric encoders use a light source and a detector with disks or plates of transparent and opaque windows. A switching operation is used which is similar to that in conducting encoders except the switching is accomplished by breaking the path of the light beam between the source and detector. These optical encoders offered improvement over mechanical types since the switching is done by means of a light beam. The transparent and opaque windows depict either the relative or absolute position of a rotary or linear shaft.

The early optical encoders used a hot filament bulb for the source with or without a lens, and the light detector was an open semiconductor chip. The hot filament bulb was subject to premature failure due to vibration and voltage surges. Most of these units were designed for easy bulb replacement and were used for noncritical applications. Accuracy was a function of brightness, which increased the size and power of the bulb and reduced reliability.

The more recent use of light-emitting diodes (LEDs) provides a light source which is not damaged by vibration and has a life in excess of several hundred thousand hours. Power consumption as well as size were greatly reduced. These features allowed linear encoders to be developed as an alternative to rotary shaft encoders used with rack and pinion gearing (Fig. 8-6).

Precision encoders can be of the incremental type, which use one or two tracks of equal spaces of transparent and opaque windows to produce a series of pulses as a function of position. The direction is detected by comparing the two outputs.

In an incremental encoder a series of position-indicating pulses is created as the light path is interrupted. These must be stored and subtracted from the number of pulses which indicate the starting position to obtain the relative position as the rest point. Counting and storing the position pulses is done with electronics either internal or external to the transducer enclosure.

An absolute position encoder requires no signal processing to determine linear position and utilizes a multiple track and arrays of emitters and detectors as shown in Figure 8-7.

The laser interferometer is a high-precision measurement unit that uses a laser beam which is directed at a reflector on the measured object. A change in the linear displacement of the object produces interference fringes which are counted

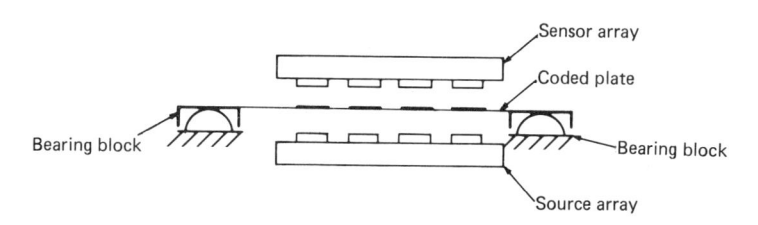

Figure 8-6. Linear optical encoder.

190 TRANSDUCERS FOR AUTOMATION

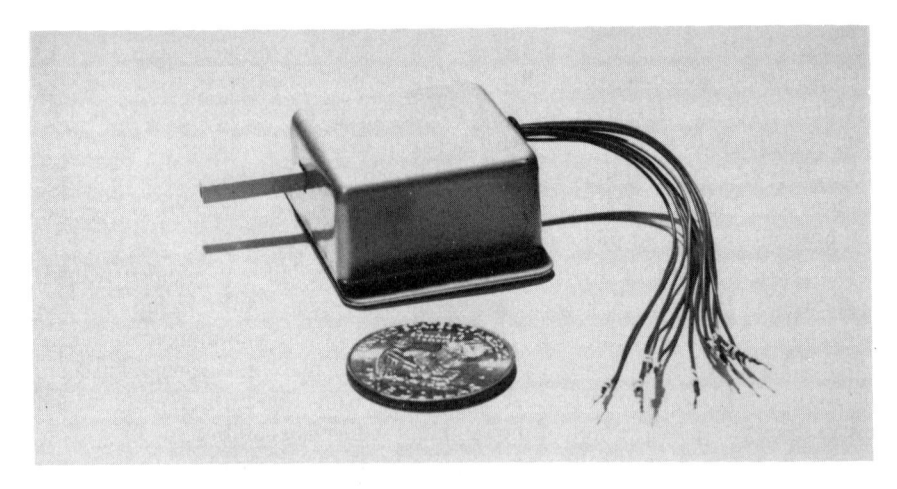

Figure 8-7. An absolute-position encoder with an 8-bit output for interfacing with a microprocessor system. (*Courtesy Siltran Digital*)

by electronic circuitry. The system is very accurate, but expensive and too complex for most automation applications.

POSITION TRANSDUCER CONSIDERATIONS

Since there are a variety of displacement and proximity sensors to choose from, selection should begin with the technical and economic requirements. The primary criteria are the characteristics of the system in which the sensor is to be used. This determines the type of output needed.

The next most important criteria involving selection are the range and overall system accuracy, which tend to reduce the number of choices. One must also consider the physical requirements of the system, since either coupled or noncontacting sensors might be used.

In dc systems potentiometric transducers tend to be simple to apply and can be used with output levels to 50 V or higher. Devices are available for displacements to 24 in. or from 5° and 3600° using multiturn units. The resolution is approximately 0.5% of full scale for the smaller displacements.

Increasing the displacement allows a greater number of wire turns with lower resistance to be used in the wire-wound devices. Accuracy and friction error tend to improve with the longer-range devices.

The reluctive transducers with dc to dc conversion circuitry are also useful in dc applications. The displacement ranges between 0.01 and 120 in. for the linear devices and between 0.05° and 90° for rotary devices. Full-scale output is typically adjusted to 5 V; however, lower and higher outputs are available.

Capacitive and inductive proximity sensors as well as photoelectric sensors may be used for the measurement of small displacements. These noncontacting sensors can be used to detect displacement changes as small as one microinch. Typical automation applications of noncontacting proximity sensors include the measurement of shaft eccentricity, bearing film thickness, rolled sheet thickness, and machined parts.

In ac systems, multiple-coil inductive sensors have been widely used. The displacement ranges are the same as for the dc output versions. The output is proportional to the excitation voltage and to the excitation frequency, which ranges from 50 Hz to 10,000 Hz. Single-coil inductive as well as potentiometric sensors have also been used in ac measuring systems.

Digital control systems can achieve a high system accuracy with incremental and absolute digital displacement sensors along with units such as interferometers. Some shaft angle encoders can provide a system accuracy of less than five seconds of arc.

The total cost of a digital system may be greater than that of an analog system, but a digital system is adaptable to microcomputers or displays with less software. They are the logical choice for most automation position control systems. Digital encoders are used in applications like pulp and paper manufacturing and automatic plotting and drafting machines.

Other techniques that are used for thickness measurements in manufacturing are listed in Table 8.1. Some of these are illustrated in Figure 8-8.

LINEAR VELOCITY MEASUREMENT

Linear velocity transducers are usually based on electromagnetics, using a change in flux to induce an electromotive force in a conductor. The flux change is produced by the relative motion between a coil and a permanent magnet. Some of these transducers use a fixed magnet with the moving coil as the sensing

TABLE 8.1. Techniques used for thickness measurements in manufacturing.

Technique	Principle	Sensitivity	Applications
Differential or single roller	Rolling contact	20 μ in.	Foil, sheets, film
Sonic	Resonant vibration	0.01 in.	Rigid sheets, accessible from only one side
Radiation	X-ray attenuation	0.01–0.001 in.	Metal foil, plastic film
Capacitance	Dielectric thickness	0.001 in.	Insulating sheets, film

192 TRANSDUCERS FOR AUTOMATION

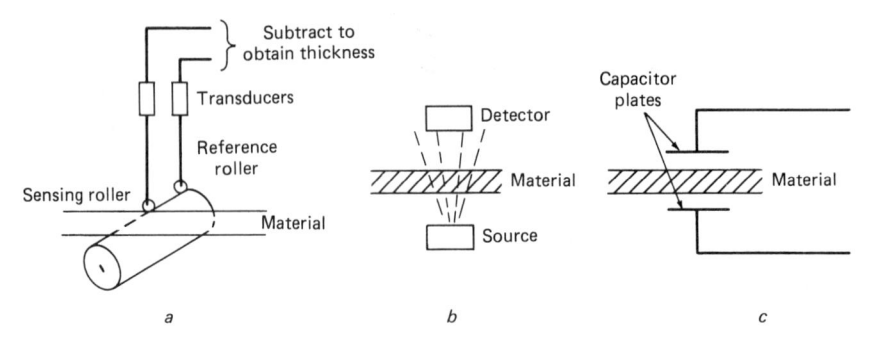

Figure 8-8. Thickness measurement techniques.

element and others use a coil that is fixed while the moving magnet is the sensing element.

The electromagnetic type of linear velocity transducer consists of a coil in a steel housing and a cylindrical permanent magnet or core with a threaded end as shown in Figure 8-9. The core moves freely with the motion of the object to which the shaft is connected. The instantaneous output voltage of the coil is proportional to the velocity of the core.

Another type of fixed-coil velocity transducer uses a permanent magnet supported between two springs. Bearings on the ends of the cylindrical magnet allow its motion within a stainless-steel sleeve with a minimum of friction. The mechanical assembly allows relative motion between the coil and magnet which causes the flux change necessary to obtain an output voltage that is proportional to velocity.

In the moving-coil transducers the coil is part of an armature which is pivoted using bearings at the end of the transducer assembly. The coil moves within the magnetic field established by the pole pieces of a fixed magnet. Linear velocity may also be obtained with a gear rack and electromagnetic sensor as used for position measurement. (See Fig. 8-5.)

TACHOMETERS

Angular speed sensors may use an electromechanical tachometer technique. The actual sensing shaft may be a solid cylinder or a radial hole that is either splined, serrated, square, slotted, threaded, or conical. In some noncontacting angular

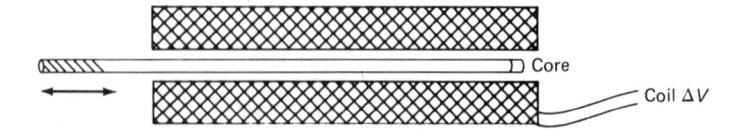

Figure 8-9. Electromagnetic linear velocity transducer.

speed tachometers a rotating member other than a shaft acts as the sensing element. These electromagnetic angular speed transducers have an output that is a varying dc voltage or an ac voltage varying in either amplitude or frequency.

Direct-current generator/tachometers (Fig. 8-10a) use a permanent dc magnet or a separately excited winding as the stator and a conventional generator with a commutator-equipped rotor. The output of the stator winding type of unit is between 10 and 20 V at 1000 rpm. The brushes needed by the commutator require periodic replacement maintenance, but an advantage for some automation applications is the indication of direction of the shaft rotation since the output may depend on it. This feature makes the dc tachometer an angular velocity transducer.

Alternating-current induction tachometers (Fig. 8-10b) operate as variable coupling transformers in which the coupling coefficient is proportional to the speed of the shaft. As the input winding is excited by an ac voltage, an ac voltage appears at the output terminals of the secondary winding. The amplitude of this voltage varies with rotor speed.

A squirrel cage type of rotor is used in some devices while others use a simple cup-shaped rotor made of a high-conductance metal. The shaft rotation produces a shaft of flux distribution on which the operation is based.

Alternating-current permanent magnet tachometers use the magnetic interaction of a permanent magnetic rotor and a stator winding to provide an ac output. The amplitude as well as the frequency of this ac magneto type of unit are proportional to the angular speed of the shaft.

For dc applications circuitry can be used to convert the output into a dc voltage. An advantage of operating on the frequency output instead of the amplitude, is the relative freedom from the effects of loading, temperature variations, and armature misalignments that may be caused by vibration.

A toothed-rotor tachometer uses a ferromagnetic toothed rotor with the transduction coil wound around a permanent magnet. These are one of the most common type of frequency output angular speed transducers. The basic technique is shown in Figure 8-11.

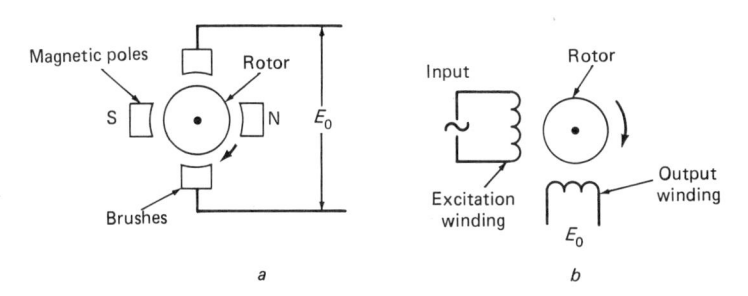

Figure 8-10. Tachometer generators. a. Direct current. b. Alternating current.

194 TRANSDUCERS FOR AUTOMATION

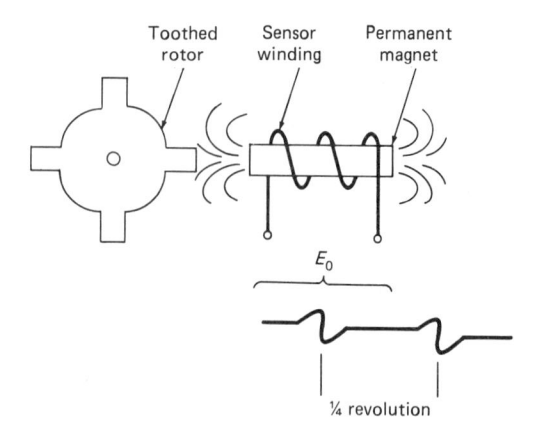

Figure 8-11. Electromagnetic frequency tachometer.

When the magnetic steel tooth on the rotating shaft passes by the permanent magnet field, the lines of flux of the magnet cut across the coil winding, inducing an electromotive force in the coil. An output pulse is created by each tooth.

The time between pulses is an inverse function of the shaft speed and can be displayed on an EPUT (events per unit time) scale. Rotors with continuous teeth are more common than single-tooth rotors since they produce a higher output frequency as well as a greater voltage amplitude at low to medium shaft speeds.

At a shaft speed of 600 rpm the output frequency of a toothed-rotor transducer with four teeth (Fig. 8-11) will be 40 pulses per second. The rms output voltage is a direct function of the clearance between the coil assembly and the teeth as well as the shaft speed.

Incremental optical encoders can also be used as tachometers. These encoders may be packaged as self-contained units or kits which are then attached to the rotating shaft that is to be monitored. The output typically has a square or a sinusoidal waveshape of constant amplitude between 2 and 8 V zero to peak.

The output frequency is a function of the shaft speed and the number of transparent sectors in the disk. These devices allow a simple interface to microcomputers in automation applications.

Reluctive angular speed transducers also use toothed rotors, but a reluctive rather than an electromagnetic transduction element is utilized. A C-core transformer with an excitation winding on one end of the core and an output winding on the other end is used in some units. A toothed rotor in the gap of the core changes the reluctance path due to the movement of ferromagnetic material in the gap.

Differential-transformer-based units use an E-core with the excitation primary winding on the center portion and two secondary windings on the outer legs of the core. A toothed rotor in one of the two gaps varies the reluctance path between the primary and one secondary winding. This type of transducer has rarely been manufactured in the United States.

Strain gauge tachometer devices use a deflecting beam with an eccentric disk attached to the sensing shaft as shown in Figure 8-12. A sinusoidal output may be obtained from the strain gauge bridge circuit, which has two arms on the bending beam. Transducers of this type have been generally produced in countries other than the United States.

Some switching angular speed tachometers use rotary switches to provide a frequency output, but it is more common for commercial units to incorporate a differentiating circuit so that a low-cost damped milliammeter can be used. This is shown in Figure 8-13. Here, the rotating contacts on each end of the capacitor can allow it to switch its polarity; R_m is the internal resistance of the milliammeter and R_w is the resistance of the interconnecting wiring if the meter is located remotely.

The meter must be sufficiently damped that it responds to the average of the individual current pulses. The average current is $kCV\omega$, where k is a constant and ω is the angular speed.

Some transducers of this type use a double-pole throw reed switch for the reversal of the capacitor polarity. Another type obtains the polarity reversal using an eccentric pin attached to the sensing shaft which moves an arm up and down to actuate the polarity contacts. Shunt resistors across the indicating milliam-

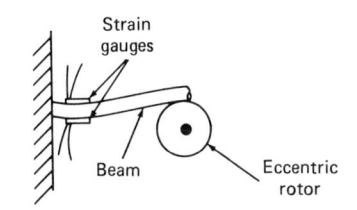

Figure 8-12. Strain gauge tachometer.

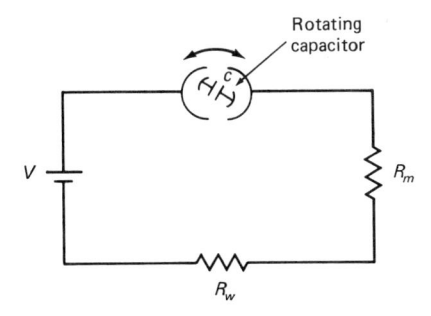

Figure 8-13. Switch-type differentiating tachometer.

196 TRANSDUCERS FOR AUTOMATION

meter are used to adjust the meter range. Other reed switching designs are magnetically actuated as shown in Figure 8-14.

ACCELERATION TRANSDUCERS

Accleration transducers use a seismic or proof mass, which is restrained by a spring system. The motion of the seismic mass is usually damped as shown in Figure 8-15. As an acceleration is applied to this type of transducer, the mass moves relative to the transducer case or frame. When the acceleration ends, the spring returns the mass to its original position. If an acceleration is applied to the transducer in the opposite direction, the spring is compressed.

Under steady-state acceleration conditions, the displacement of the mass is given by the acceleration multiplied by the ratio of the mass M to the spring constant k. Under dynamic or varying acceleration conditions the damping constant enters into a modified version of this equation.

The seismic mass in a linear accelerometer is usually a circular or rectangular section. It may be linked to the case by slides or bars, but it is always restrained from motion in any but the sensing axis.

The seismic mass of an angular accelerometer may be a disk pivoted at its center and restrained by a spiral spring. It responds to angular acceleration with an angular displacement.

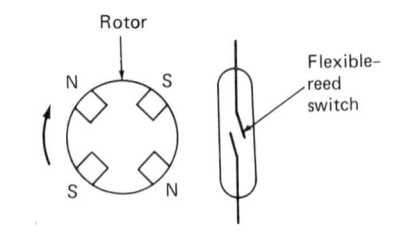

Figure 8-14. Magnetic-reed switch tachometer.

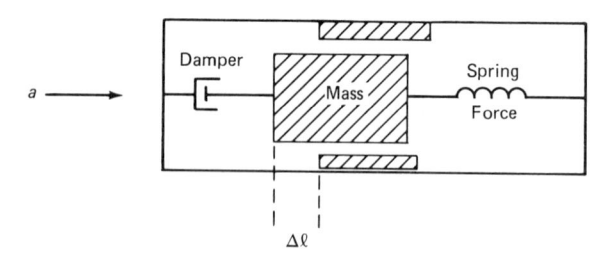

Figure 8-15. Basic acceleration transducer.

Capacitive acceleration transducers use a change of capacitance in response to acceleration. Some designs use a fixed stator plate and a diaphragm to which a disk-shaped seismic mass is attached. The diaphragm acts as the restraining spring as well as the moving electrode of the capacitor. An acceleration acting on the mass causes the diaphragm to deflect and its capacitance with the stator changes proportionally.

PIEZOELECTRIC ACCELEROMETERS

Piezoelectric accelerometers use the force on a crystal to measure acceleration. The crystal can be bonded to the mass, which also may act as the elastic member. Some designs use an annular crystal which is bonded to a center post on its inside surface and to an annular mass on its outside surface. The upward or downward deflection of the mass causes shear stresses across the thickness of the crystal.

Most units designed for compression use a stacked arrangement in order to increase the low output voltage of the quartz crystals. A typical transducer may use six or seven crystals which are stacked and connected as shown in Figure 8-16. The impedance across the crystal electrodes is generally high. A coaxial connector is typically used for all piezoelectric acceleration transducers.

The base and case of many piezoelectric accelerometers are connected to one crystal electrode or set of electrodes to reduce noise. The case is also sealed to prevent the entry of moisture, which could affect the output.

The larger accelerometers tend to have a higher sensitivity and lower frequency. The higher the resonant frequency, the lower the capacitance or sensitivity and the more difficult it is to provide mechanical damping.

The amplification factor of these transducers is defined as the ratio of the voltage sensitivity at its resonant frequency to the voltage sensitivity in the band of frequencies in which the sensitivity is independent of frequency. This ratio is indirectly proportional to the amount of damping in the seismic system. Some transducers with resonant frequencies below 20 kHz use silicon oil damping but

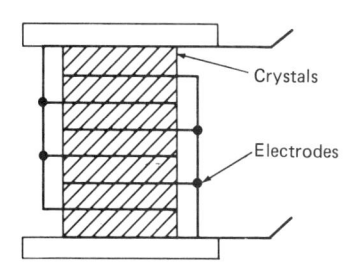

Figure 8-16. Stacked-crystal piezoelectric accelerometer.

198 TRANSDUCERS FOR AUTOMATION

most of the piezoelectric devices are essentially undamped and have amplification ratios in the range of 5 to 50.

ACCELEROMETER MEASUREMENT SYSTEMS

The total measurement system capacitance includes the cable capacitance, so the cable length can be critical. For the maximum frequency and voltage output, source followers can be used for impedance matching. The use of followers can solve most of the problems associated with high-impedance voltage amplifiers.

Calculations can be made for a particular source follower to determine the frequency and output voltage for different values of cable length, or the maximum output voltage and/or frequency for a given cable length. The problem of matching with a follower can be minimized by the use of a charge amplifier system.

A charge amplifier is an operational amplifier with capacitive feedback as shown in Figure 8-17. This circuit allows E_{out} to be proportional to the charge produced by the accelerometer.

This type of amplifier detects charge rather than voltage so the measurement system is independent of shunt capacitance and insensitive to any variations in shunt capacitance. Thus, the cable length can be ignored in the measurement system output calculations and the system characteristics of the transducer are not affected by changes in capacitance of the transducer or the cable. Other advantages of the charge amplifier approach include a lower dynamic input impedance and a flat freqeuency response to 10 Hz or less. The dynamic impedance is usually less than 1 megohm and this reduces noise as well as the effects of humidity and cable connector contamination.

The charge amplifier can provide an output voltage which closely corresponds to the actual motion of the accelerometer mass. The amplifier output is a function of the transducer charge sensitivity, the actual vibration amplitude, and the amplifier gain setting. The actual waveform of the signal is dependent on the frequency response of the amplifier. The temperature response of the system will depend on the charge characteristics of the accelerometer. Temperature

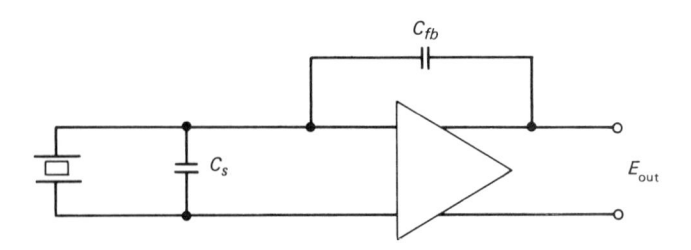

Figure 8-17. Charge amplifier.

DISPLACEMENT AND PROXIMITY MEASUREMENT 199

compensation is usually accomplished with a series capacitor. This will flatten the charge versus temperature curve but it also affects the gain of the charge amplifier.

Acclerometer sensitivity can be determined by mounting it on a vibration table. The table is set to the desired peak-to-peak displacement, for example, 0.050 in. at 100 Hz, and the displacement is measured optically while the frequency is monitored by a counter. The acceleration level may be determined to an accuracy of $\pm 1\%$ from

$$a = 0.1023 df^2,$$

where

a = resultant acceleration in peak g,
d = peak displacement,
f = frequency.

The charge sensitivity can be found using

$$S_c = \frac{(2VC_t)^{1/2}}{ka},$$

where

S_c = charge sensitivity in coulombs per gravitation unit,
V = system output in volts rms,
C_t = total accelerometer, cable, and follower capacitance, in picofarads;
k = follower gain;
a = acceleration level in peak g.

The open-circuit voltage sensitivity can then be determined from

$$S_{OV} = \frac{S \times 10^3}{C},$$

where

S_{OV} = open-circuit voltage sensitivity,
C = accelerometer capacitance in picofarads.

The cross-axis sensitivity or traverse response can be found by mounting the accelerometer on the sensitive axis and recording the output at a particular frequency and displacement. The accelerometer is then mounted perpendicular to the sensitive axis and rotated 360° while vibrating at the same g level to determine the maximum output. The cross-axis sensitivity is the maximum output perpendicular to the sensitive axis divided by the output on the sensitive axis at the fixed vibration level.

The frequency response of an accelerometer can be found by using a calibrated shaker. The drive of the table is calibrated by using a certified standard, then the response is read on a working standard. The certified standard is then replaced by the accelerometer to be tested. The data from the standard are averaged while the working standard is used as the table drive control during the tests.

The natural frequency can be determined by mounting the accelerometer on a plate that has at least ten times the effective mass of the accelerometer. Then this system is suspended and a variable-frequency oscillator is used to drive the accelerometer. The accelerometer current is held constant as the frequency is changed until the output across a resistor is a maximum with a 90° phase shift. This is the natural frequency of the accelerometer plate system as well as the frequency of minimum impedance.

The temperature response may be found by measuring the accelerometer output at a particular frequency and level such as 80 Hz and 2.5 g at the ambient temperature and then at the temperatures of interest.

EXERCISES

1. Explain the difference between a displacement sensor and a proximity sensor using some examples.
2. A linear potentiometer is to be used as a displacement sensor. It has 5 V excitation across it and is made up of 1200 turns of wire of 0.0038-in. diameter. What is the expected noise voltage amplitude?
3. An angular position transducer uses a rotary potentiometer with a conductive film element. The contact resistance changes by 0.1 ohms. The external load is 2 kohms. The transducer is a 100-ohm unit with 5 V across it. Find the noise voltage generated by the potentiometer.
4. Consider the operation of a capacitive sensing circuit and discuss the most critical parameters.
5. Discuss the effects of the excitation frequency on differential transformer performance.
6. A differential transformer has a sensitivity of 5 V/in. If this transducer senses a movement of 0.0015 in., what will the output be?
7. Show the operation of an optical encoder using a diagram. What are the critical parameters that affect the output level?
8. What is the difference between an absolute and an incremental encoder? How does this affect the interface to a microcomputer?
9. Draw the plate pattern for a four-bit linear binary conducting encoder. What would be the advantage of using a cyclic binary code or a BCD coded plate?

DISPLACEMENT AND PROXIMITY MEASUREMENT 201

10. It is desired to monitor the thickness of plywood sheets during manufacturing. Select a transducer and show the complete system block diagram of a microcomputer monitor and control system for the thickness and pressure controls required. The microcomputer will control the hydraulic pressure to the plywood pressure rollers.
11. Discuss some basic techniques used to measure linear velocity.
12. What are the advantages of an optical encoder tachometer over a reluctive or strain gauge tachometer in a microcomputer control system?
13. What are the critical parameters when a linear gear and electromagnetic sensor are used to measure velocity?
14. Describe the differences between dc and ac tachometers.
15. Describe the characteristics common to all accelerometers.
16. What are some of the characteristics of the piezoelectric accelerometer? Write an equation defining the amplification factor.
17. What is the advantage of using a charge amplifier in accelerometer measurement circuits? What are its drawbacks?
18. Write an expression defining the cross-axis sensitivity for an accelerometer.
19. It is desired to measure the temperature response of piezoelectric accelerometers as a quality control test. Write a test procedure and select a temperature monitoring method. The data are to be collected and analyzed by a microcomputer system. Define the operation required with a flow diagram.
20. Discuss a system for automating the calculation of the natural frequency of piezoelectric accelerometers. The accelerometers weigh four ounces and it is desired to sort 200 per hour. Define the basic equipment needed and draw a flow diagram.

9
OTHER PHYSICAL MEASUREMENTS

TURBIDITY

Several of the basic measurements discussed above have to do with the amount or quantity of a material. Flow describes the rate of movement, while accumulation is measured by pressure or level. Temperature has to do with the condition or quality of the material. In some cases, temperature completely describes the condition of the material, as with the temperature of boiling or freezing water under ambient conditions.

Temperature alone is inadequate to describe the condition of some complex chemicals, or mixtures. For example, plant effluent or waste discharge may be fully described by four measurements—pH, conductivity, dissolved oxygen, and temperature. Other materials may require additional measurements.

Some of these include density and moisture measurement, as examples. Instruments have become available, in many cases quite recently, for the measurement of many physical and chemical properties. Most of these sensors have been adapted from laboratory instruments—early units were too fragile for use in an operating environment and continuous maintenance was required. More rugged versions were developed, and these sensors can be expected to operate reliably for extended periods with only minimum protection from service conditions that could cause failure.

Turbidity is one of these parameters. It is an optical property of liquids caused by the presence of suspended particles. The particles cause a scattering of the light energy passing through the liquid, and the turbidity is influenced by the concentration, size, shape, and optical properties of the particles in addition to the optical properties of the fluid.

The basic theory of the scattering of light by particles was developed late in the nineteenth century. The theoretical understanding of the light scattering by particles is well established, but the general theories are complex and require computers for many analytical solutions. However, this complexity is seldom required for the measurement of turbidity in automation applications.

The Jackson candle turbidimeter and the Jackson turbidity unit are the standard instrument and unit of turbidity measurement. The Jackson turbidimeter comprises a special candle and a flat-bottomed glass tube graduated in Jackson turbidity units. The turbid sample is poured into the tube as the image of the candle is observed from the top. When the image disappears into a uniform glow, a reading on the depth of the sample is made.

The Jackson turbidimeter compares the strength of the transmitted light with that scattered or reflected. The yellow red candle flame is at the long-wavelength end of the visible spectrum, and it is not effectively absorbed or scattered by fine particles. Thus, the candle flame will not disappear, and the Jackson turbidimeter cannot be used for such particles. Black particles tend to absorb all of the light, and the liquid becomes dark before enough sample can be poured into the tube to reach the image extinction point.

These difficulties were overcome in modern instruments using incandescent light sources which provide a wider light spectrum, or by using comparison rather than extinction techniques for measurement. The nephelometer is a turbidimeter which measures the right-angle light scatter by a fluid, and is sensitive to low turbidities.

The absorbometer is a turbidimeter which measures light passing through the sample. It is insensitive to small turbidities, but has no upper limit on the turbidity range. If two beams of light from a single source are monitored by these two techniques and the results compared, a much wider range of measurement can be obtained.

Most modern turbidimeters use a voltage regulated light source mounted on one side of a sample chamber or cell C (Fig. 9-1). The light projects a beam of energy through the liquid sample, and as the energy is projected through the fluid, a part of the radiant energy is deflected or absorbed by the particulate matter. The radiant energy reaching the far side of the sample chamber is detected by a photometric detector, usually a photocell or bolometer, which produces an electrical signal which is a function of the energy detected. The detector is usually one arm of a Wheatstone bridge.

THERMAL CONDUCTIVITY

Thermal conductivity is well understood theoretically, but in practice, accurate measurements can be difficult to complete. Both steady-state and dynamic thermal conductivity measurements can be made.

Steady-state measurements are based on the one-dimensional equation for

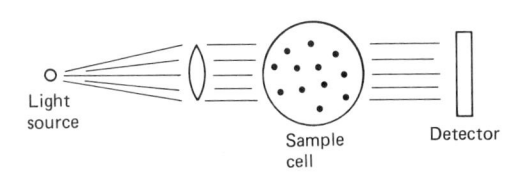

Figure 9-1. Optical turbidimeter.

conductive heat transfer. Steady-state measurements of conductivity can be made by measuring (1) the power dissipation in an electrical heater, (2) the temperature in a constant flow of water, (3) the boiling off of a liquid of known thermal properties.

One method of steady-state measurement uses a guarded hot plate. Here, two samples are placed on either side of a heater or hot plate assembly. The heater assembly consists of an inner heater surrounded by an annular outer or guard heater.

The guard heater eliminates radial heat flow from the inner heater, resulting in one-dimensional heat flow through the sample. Heat sinks on the opposite sides of the sample are used to provide a temperature gradient across the sample. Thermocouples are installed on each surface of the sample. The temperature difference across the sample then provides a measure of the thermal conductivity.

The thermal conductivity gas analyzer can be used for dynamic measurements. It consists of two wires, usually platinum, each of which is a leg of a Wheatstone bridge circuit. Each wire is stretched through a measuring cell.

A reference gas flows through one measuring cell, while the gas sample flows through the second cell. One filament wire is exposed to the reference gas, and the other to the sample gas. This allows a comparative measurement between the reference and sample gas.

Since thermal conductivity is the ability of a fluid to transfer heat, a gas that transfers heat from a source at a greater rate than another gas has a higher thermal conductivity. The dual conductivity cell allows a comparison of the thermal conductivity of an unknown gas to that of the known reference gas.

The wire filament in each side of the cell is heated to a low temperature by an electrical current. Heat is lost from the filament to the gas in each side of the cell. The heat loss varies with the conductivity of the gas and changes the filament temperature and resistance. The resistance is then measured with the bridge circuit by a comparison to the resistance of the filament in a reference gas of known conductivity.

A conductivity analysis may be made to identify the volumetric percentage of one gas in a gaseous mixture. The proper identification and calibration is made possible by the choice of the reference gas.

Non-steady-state methods are used to determine diffusivity, which is the ratio of thermal conductivity to specific heat. As the diffusivity is being measured, the dynamics of the conductivity measurement vary with the gas composition. A response time of 5 s is typical for a hydrogen measurement, while 25 s is typical for carbon dioxide.

The resistance measurement tends to drift with time because of the vaporization of the filament, and there may be some reaction of the hot filament with the gas. The filament is usually coated with glass to eliminate drift, but this coating also tends to slow the response of the measurement.

ELECTRICAL CONDUCTIVITY

Conductivity measurement is the determination of a solution's ability to conduct an electric current. Pure water is a nonconductor, and has a conductivity of essentially zero. When an electrolyte (a material that ionizes such as an acid or a base) is dissolved in water, the solution will have a unique conductivity, depending upon the electrolyte, the concentration, and the temperature of the solution.

A measurement of the conductivity is thus a measurement of the concentration of the electrolyte in the solution at the given temperature. Table 9.1 show the variation of the conductance of a salt versus the concentration of the solution.

The coefficient can vary greatly, depending on the nature and concentration of both the electrolyte and the non-ionic materials in the solution. Increasing temperature almost always increases conductivity, and thus most conductivity measurements must be temperature compensated. Temperature compensation means that the compensated reading is the reading that would be obtained if the solution were at some reference temperature, usually chosen as 25°C.

The earliest attempts to use solution conductivity in industrial measurements occurred in 1910 in the determination of the purity of distilled water. These early sensors were based on fragile laboratory devices, but continuous developments since then have resulted in sturdy and stable instruments which are able to withstand the demanding environments of process streams.

The use of stainless-steel and alumina materials resulted in a rugged device that can be used for pressures to 10,000 psi. Removal techniques such as packing and gate valves allow the cell to be removed and reinserted into a process stream under line pressure and temperature conditions.

Conductivity is usually measured by immersing two conducting electrodes in a liquid and applying an ac voltage across the electrodes. Alternating current is required since direct current will cause polarization of the electrodes and formation of a high resistance at the electrode surfaces. The high resistance will then alter the conductivity measurement. The ac voltage forces a current to flow between the electrodes.

TABLE 9.1. Conductance of water.

	Conductance (mho/cm)
Ultrapure water	0.05×10^{-6}
Distilled water	1×10^{-6}
Rain water	50×10^{-6}
0.05% salt solution	1×10^{-3}
Sea water	20×10^{-3}

The resistance provided by the ionized solution to the electrical current between the electrodes is usually measured by a Wheatstone bridge circuit, where it is converted to units of conductivity.

Specific conductivity is defined as the reciprocal of the resistance in ohms measured between opposite faces of a centimeter cube of a solution at a specified temperature. The cube is made up of the area of the electrodes multiplied by the distance between them, as shown in Figure 9-2. The conductivity is proportional to the cross-sectional area of the cube, and inversely proportional to the distance (d) across it.

Conductivity, like viscosity, turbidity, and thermal conductivity, is an inferential measurement; it carries no information about the ionic carriers of the current. Thus, a measurement of concentration requires that the conductivity-concentration (Fig. 9-3) be known. For this reason, most industrial applications of conductivity are in the area of monitoring water purity. This includes quality measurements for feed water and the detection of contamination due to leaks in process coolers. The use of these measurements extends to determining the

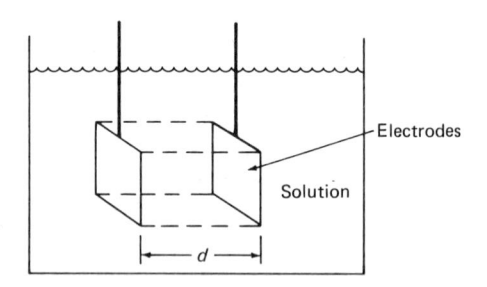

Figure 9-2. Conductivity measurement.

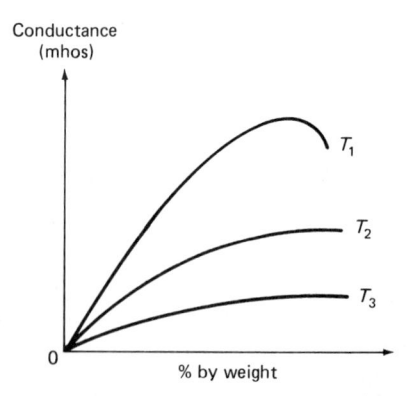

Figure 9-3. Conductivity concentration.

completeness of reactions involving ionizable materials, and the mixing effects in solutions.

Care must be exercised in the monitoring of electrolytes dissolved in water. The likelihood of errors is especially high when dealing with relatively low concentrations of unbuffered solutions. This is due to the extreme mobility of the hydrogen ion, which causes relatively small changes in pH to affect the conductivity greatly. The calibration curves relating conductivity to concentration then become questionable.

HYDROGEN ION CONCENTRATION

The concept of pH involves the measurement of the acidity or alkalinity of a liquid which contains a proportion of water. The following equation shows the pH to be related to the concentration or the strength of the liquid solution. This concept of pH was first introduced in 1909 by S. Sorenson.

$$pH = \log C,$$

where

$$C = \text{concentration}.$$

This relationship permits the assignment of an exact value of pH to a specific concentration. The advantages of this relationship are well known in the commercial manufacture of many products, for example, in the food industry.

Many products tend to be difficult to produce consistently without the proper control of pH. An example is products that are required to jell at specific temperatures and pressures. Some batches may be too thick or too thin, and others may fail to jell at all. Usually the following type of relationship between the product characteristics and pH will hold:

pH	Condition of product
below 2.5	Will not jell
2.8	Maximum stiffness
3.0	Medium stiffness
above 3.2	Will not jell

The measurement of pH thus allows the manufacture of uniformly consistent product with the desired characteristics.

Soon after Sorenson originally defined pH, G. Lewis suggested that ionic activity is separate from component concentration, and that the pH galvanic cell measures ionic activity rather than concentration. The definition then becomes

$$pH = -\log C_H,$$

where

$$C_H = \text{hydrogen ion concentration.}$$

The ion is the charged particle that determines the current-carrying capacity of the solution. The measurement of pH is thus the measurement of the dissociation of the acid or alkali molecules into ions. Some acids dissociate readily, and produce high concentrations of hydrogen ions. These are known as the strong acids such as sulfuric acid. Boric acid, which can be safely used as an eyewash, is a weak acid and does not dissociate readily.

The measurement of pH is meaningful only for dilute concentrations of an acid or alkali dissolved in water. This is due to the fact that only at high dilutions does the activity coefficient, which is the relationship between hydrogen ion concentration and component concentration in the two definitions, approach unity.

The dissociation of water molecules controls all reactions in aqueous solutions. Water dissociates slightly, since it is a weak electrolyte, into equal numbers of hydrogen (acid) and hydroxyl (alkali) ions. Defined as neutral, pure water has a pH of 7.

One common method of pH measurement is based on the use of substances which change color when subject to acidic or alkaline environments. If litmus paper is dipped into the liquid sample, it changes to a different color depending on whether the sample is acidic or alkaline. Quantitative measurements require the use of liquid indicators which change color at a specific pH. The concentration of hydrogen ions in the solution changes the constituents of the indicator, which affects the color.

Measurement of pH to within 0.1 pH units is possible using these liquids in the laboratory. The major difficulties associated with the method are

1. Accuracy is affected by the color of the sample.
2. Addition of the indicator may affect the pH value of the sample.
3. The measurement if affected by dissolved or suspended matter in the sample.

The technique used for the measurement of pH in the processing industries depends on measuring the voltage developed by two electrodes in contact with the electrolytic solution. The measuring system consists of three major elements: the measuring electrode, the reference electrode, and electronics to sense and amplify the small voltage developed between the two electrodes.

The measuring electrode produces a dc voltage which is proportional to the pH of the stream being measured. The electrode surface is a membrane which responds to the hydrogen and hydroxyl ions with an internal exchange of ions within the membrane. The membrane is fused to a glass body such that the outer surface makes contact with the process fluid while the inner surface is in contact with an internal filling solution. The internal electrical connection is made to the glass surface using a buffer solution, which is typically in contact with a couple surrounding the lead wire.

The reference electrode is inserted into the stream where pH is being measured and it becomes the second conductor to complete the electrical circuit. It is insensitive to all ions and can be compared to the cold reference junction of a thermocouple circuit.

One type of reference electrode uses a silver–silver chloride couple to make contact with a saturated potassium chloride solution which is allowed to flow through a small opening to the process stream. The silver–silver chloride electrode is a silver wire which is coated with silver chloride. The electrical contact between the potassium chloride solution and the process is through the porous surface of the electrode.

The measurement circuit must be capable of measuring voltage potentials in the millivolt range and must have a high input resistance. This is necessary to minimize current flow, since the current flow multiplied by the electrode resistance produces a voltage divider and presents an error in measuring the generated voltage.

The necessary flow through the porous plug in the reference electrode which is required to complete the electrical circuit has proved to be a problem in the past. Flow stoppages due to coating of the porous plug or insufficient pressure within the electrode to maintain the flow can require frequent cleaning.

Solid-state electrodes require no electrolyte flow. Potassium chloride and silver chloride crystals are sealed in a cell, and the process liquid enters the cell through a wooden or ceramic porous plug. The liquid enters the reference chamber and dissolves some of the salt to form a conductive path between the silver–silver chloride half cell and the liquid being measured. The porous plug can still become fouled but it has substantially more surface area and is not blocked as easily. The characteristics of pH sensors are given in Table 9.2.

FLAME SENSORS

Flame sensor applications fall into three categories. Checking the pilot and main flames simultaneously is the most desirable method of flame supervision. This

TABLE 9.2. pH sensor characteristics.

Pressure	to 150 psig (1 MPa)
Temperature	23 to 212°F (−5 to 100°C)
Error	±0.02 pH units
Range	0 to 14 pH

allows the pilot to safely ignite the main flame. The flame rod is located in the path of both the main flame and the pilot flame (Fig. 9-4.) In some installations it may be difficult to test the pilot and main flames simultaneously because of variations of the main flame envelope at different firing rates. This often occurs in forced-air burners. In these cases the usual practice is to supervise the pilot flame. In applications where the pilot flame is not lit continuously, the main flame is supervised.

The more widely used means of flame detection depend on the following characteristics of the flame: (1) the heat generated; (2) the ability to conduct electricity (ionization); and (3) the radiation, which includes the wavelengths in the visible, infrared, and ultraviolet regions.

HEAT SENSORS

These flame sensors utilize the heat generated by the flame. Thermocouples and bimetallic elements are commonly used for small units such as domestic burners. Their relatively slow response of several minutes makes them unsuitable and often dangerous for larger units. A large industrial furnace may require 1000 scfm (0.47 m^3/s) of fuel gas. In 3 min, 3000 ft^3 (84 m^3) of unburned fuel will be stored in the furnace, creating an explosion hazard. A detector with a response of several seconds will not allow an amount of gas to enter the furnace which could cause an explosion. Another disadvantage of these sensors is that they only sense heat and cannot distinguish between heat generated by flame and that radiated by the hot refractory walls.

ELECTRIC-CONDUCTION-TYPE DETECTORS

In these sensors the detection of flame is based on the fact that a flame is capable of conducting electricity. This occurs because the flame is a chemical reaction between the fuel and oxygen and this reaction liberates a large number of electrons due to ionization.

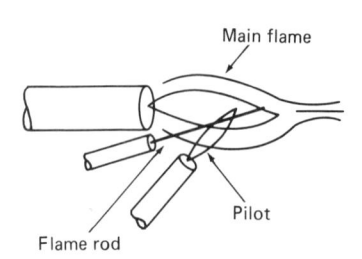

Figure 9-4. Pilot and main flame in the path of the flame rod.

The flame can conduct both direct and alternating currents, which can be utilized in an electrical circuit. A rod immersed in the flame can provide one electrode while the burner acts as the other. This is a much faster flame detection method compared to heat methods, but any path that provides a high resistance between the electrodes will be detected as the presence of flame. This false detection can be greatly reduced by using a form of current rectification. If the area of the burner is made much larger than the flame rod, the conduction will essentially occur in only one direction. As ionization takes place, electrons are liberated from the gas molecules and free to move about. If the area of one electrode is several times larger than the other, and that electrode is negative, it will accommodate a larger number of positive ions. This will also increase the flow of electrons to the other electrode, which is positive.

If alternating current is used in the flame detection system, the current through the flame will be rectified. By detecting this half-wave rectified current, the reaction to simulated flame is eliminated, since the detector circuitry will only respond to a half-wave rectified signal. An unrectified alternating current or direct-current input to the detector circuit will result in a safe shutdown.

Flame rods are in direct contact with the flame and they must be cleaned and replaced often. For applications above 2000°F (1100°C), the number of metals that can be used is limited and these can become brittle in time.

Those fuels which contain a high sulfur content produce a flame with a low resistance. This results in a low output and can cause false shutdowns. Radiation sensors can be used to avoid these problems.

RADIATION SENSORS

The radiation emitted by the flame includes wavelengths in the visible, infrared, and ultraviolet ranges (Fig. 9-5). Visible radiation represents about 8–10% of the total energy radiated by the flame. To detect these wavelengths a rectifying phototube can be used.

The rectifying phototube has a large light-sensitive cathode and a smaller anode. The anode must be positive for conduction to take place. The number of electrons emitted by the cathode is a function of the light intensity, and if ac is applied, the tube acts as a half-wave rectifier. The detector circuit is designed to respond only to a half-wave rectified signal so that high-resistance paths will not simulate a flame.

In some high-temperature applications the refractory material used to line the interior of furnaces can emit radiation in the visible range which the detector may not be able to distinguish from visible radiation emitted by the flames. This can sometimes be corrected by installing the phototube in such a way that it will be able to detect the refractory radiation.

Rectifying photocells are generally limited to oil-fired burners because the

212 TRANSDUCERS FOR AUTOMATION

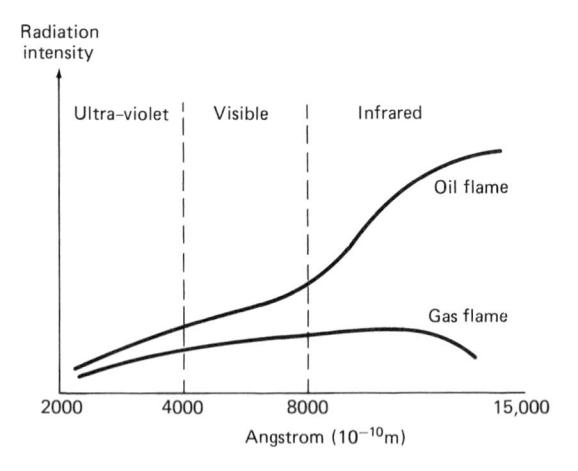

Figure 9-5. Flame characteristics.

visible radiation emitted by a gas flame is too small to be detected by this class of flame sensor.

Cadmium sulfide photocells use an element coated with cadmium sulfide, which is only sensitive to radiation in the visible spectrum. In the absence of visible light, cadmium sulfide provides a high resistance to electrical current. When it is exposed to light, the electrical resistance decreases directly with the increasing intensity of light. It conducts current equally well in either direction.

The sensitivity of the cadmium sulfide cell is such that it will not respond to gas flames, and it can only be used with oil flames. The cell is not affected by hot refractory radiation.

The infrared radiation portion includes about 90% of the wavelengths emitted by the flame, and it provides a more reliable means of detection than the visible radiation. It is emitted by gas as well as oil flames.

Lead sulfide photocells are sensitive to infrared radiation and widely used for infrared detection. They act as a variable resistor and conduct electricity in both directions without rectification. The current flow is a function of the flame strength.

A flame flickers or pulsates at a rate that is too rapid to be detected by the human eye. The frequency of this flicker is irregular but it can be detected by a circuit designed to accept a select band of frequencies. This sort of tuned circuit will not be affected by infrared radiation emitted by hot refractory materials or be affected by a high-resistance path which could simulate a flame. It is possible however for shimmering effects caused by the movement of hot gases between the refractory material and photocell to be detected as a flame under certain conditions.

The use of rectifying phototubes and lead sulfide photocells requires that the incandescence of hot refractory materials or shimmering effects due to hot air do not simulate a flame. This can be accomplished by limiting the field of view to the flame using a restricting orifice.

Ultraviolet radiation represents only about 1% of the total flame radiation energy and 10% of the emitted wavelengths. Ultraviolet (UV) radiation is present in gas as well as oil flames.

A common UV detector is a gas-filled tube with two electrodes, an anode and a cathode. If a voltage is applied between these electrodes, the tube will conduct in the presence of ultraviolet radiation. It will not be affected by hot refractory materials because ultraviolet radiation is not present in materials below 2500°F (1370°C).

Some UV detectors use a shutter arrangement that operates at a rate of about 20 times a second. This requires that the circuit test the flame 20 times a second and greatly reduces the probability of a false reading due to the shimmering effect.

In general, the heat sensors have a response that time is too slow for many automation applications. The flame rods provide for a fast response, but the direct contact with the flame causes brittleness and eventual degradation. Photocells that use visible radiation can detect radiation from refractory materials. They are also only sensitive to oil flames. The infrared detector does not suffer from these problems, but hot air shimmering effects can cause false detection. UV detectors are more reliable but have a relatively high cost. For oil-fired applications, cadmium sulfide cells are useful provided they are installed so that they will not sense hot refractory materials. In combination oil and gas furnaces, lead sulfide cells can be used if they are used with a restricting orifice to prevent false detections due to shimmering effects.

Only the flame rods are in contact with the fuel and the flame. Temperature limitations on the other devices are usually not a problem because the temperature outside of the furnace is normally below 150°F (66°C). The characteristics of various flame sensors are given in Table 9.3.

LEAK DETECTORS

Industrial materials and fuels that are not retained can represent a safety hazard as well as a financial and a resource loss. Many materials susceptible to leakage are toxic, and many are flammable or explosive.

Vessel pressurization using pneumatic or hydraulic systems is widely used.*

---

*A number of standards and codes are available for specific test procedures for piping, pneumatic instrumentation systems, and vessels. The National Fire Protection Association (NFPA) can furnish details on underground leaks and American Petroleum Institute (API) Standard 527 describes a technique for checking the seats of safety relief valves for leaks.

TABLE 9.3. Flame sensors.

	Rectifying flame rod	Phototube	Lead sulfide photocell	Cadmium sulfide photocell	Ultra-violet detector
Can be used for both gas and oil flames			×		×
Affected by refractory		×	×		
Needs shielding			×		
Limited temperature	×	×	×	×	
Flame may be simulated under certain conditions		×	×		×
High-temperature deterioration	×				

A rapid drop in vessel pressure can indicate a leak. Special equipment is available for pressure testing, including special fittings for lines and testing joints.

The bubble emission method uses a commercial formulation that tends to bubble or foam at the point of a leak. Safe materials are available for treating various chemical applications. After a leak has been detected with the pressurization method, bubble emission can help to isolate it. For detecting steam trap leaks there is a tape that is fastened to the outlet tubing. If a spot on the tape remains silver, the trap is working. If steam is getting by the trap, the spot turns black as a result of overheating. Dyes can be used in liquid streams to check flows and seepage areas. Some biodegradable, fluorescent dyes are visible after dilution to 1 ppm in water. There are a number of aerosols, paints, and papers that change color with chemical action and these are all applied to the exact point to be tested. Phosphorescent powders can be used for locating leaks in dust collectors. The powder works its way through the collector and the inside of the collector is checked with an ultraviolet light to find the leakage points, which are highlighted with the glowing powder. White smoke can be used in ducts and other places that cannot be pressurized. Candles, bombs, and generators with blowers are available to produce various quantities of smoke.

A number of different types of analyzers are available for toxic or flammable liquids. These instruments can be based on infrared, calorimetric, or other principles. Chemical or physical properties may be measured with units that range from portable to continuous, permanent monitors.

Some portable detector units allow measurements of low concentrations of gases and vapors, using the length of a discoloration in a tube to provide the concentration reading. Other tubes use chemicals that change color and require a color comparison. There are tubes for over a hundred specific types of analysis including tubes that have been certified by the National Institute for Occupational Safety and Health (NIOSH).

A common type of combustible gas detector uses catalytic combustion. A platinum wire filament, which is connected in a bridge circuit, is mounted in a diffusion sensing head. The technique is useful for the intermittent and continuous monitoring of flammable gases and vapors, although sulfur compounds and halides can impede the operation of the sensor. H_2S detectors based on semiconductor sensing elements are available in the forms shown in Figure 9-6.

A mass spectrometer that is sensitive to helium can be used for leak detection. The helium acts as a tracer gas, inside or outside of the unit under test. These instruments are expensive but they have considerable value in production leak testing.

Thermal conductivity leak detectors use sample cells that contain coils in a bridge circuit configuration. The heat dissipation depends on the concentration of the gas or gases in the sample. These instruments can be used to detect many

a

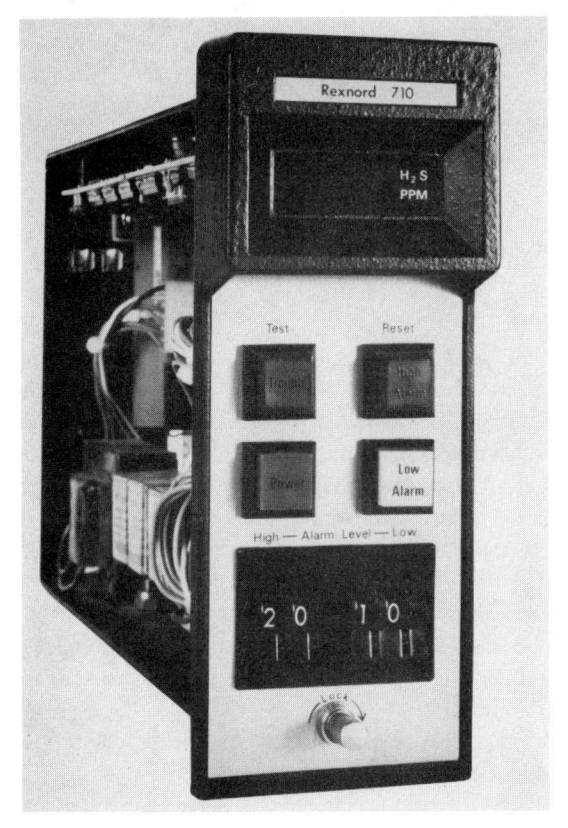

Figure 9-6. H_2S detectors. *a*. Modular unit for rack mounting. *b*. Four-channel unit. (*Courtesy Rexnord*)

216 TRANSDUCERS FOR AUTOMATION

b

Figure 9-6. (*Continued*)

different gases over a wide dynamic range. One type uses halogen to produce ionization and current flow. Although this type of instrument is useful for checking refrigeration systems and for certain kinds of production testing, it cannot be used in many industrial applications since the cell operates at too high a temperature to be used in a flammable environment.

Thermography techniques detect temperature differences using infrared scanning. This produces a thermal outline in several colors or shades. If a hot fluid, such as steam is leaking, it can be detected at a distance, underground, or through insulation.

Thermography can also be used to inspect insulation, refractory materials, heater tubes, and electrical equipment for heating.

Acoustic emission techniques are based on the detection of the sonic and ultrasonic waves generated by fluids escaping from openings. The acoustic emission technique has several advantages. Leaks can be detected at a distance and the method can be used for the location of leaks from buried pipes and tanks.

Although the technique is limited for low-pressure gas leaks, it can be useful for detecting leaks from low-pressure liquid lines.

Other leak detecting techniques include the following:

1. Flow meters can be used to measure the mass flow in and out of a system.
2. Flow-balance algorithms can be used to detect losses.
3. Optical density measurements can be used for the detection of broken bags or other leaks in dust collectors.
4. The rf power loss can be measured to detect leaking or broken packages for moisture.
5. The measurement of conductivity can be used to check the integrity of the walls of glass-lined vessels and heat-exchanger tubes. A nonconductive liquid is checked for contamination using a conductive fluid.
6. Leak detectors are available in the form of a cable that is degraded by contact with petroleum-derived materials.

METAL DETECTORS

Most metal detectors use magnetic fields. The magnetic circuit of the search coil is affected by the differential conductance of the medium, which is usually sand, earth, or water. The greater the difference between the conductance of the medium and that of the minerals or metallic objects, the greater the sensitivity of the detector and smaller objects may be detected at a greater distance.

There are two types of detectors in use, (1) the transmit-receive (T/R) type and (2) the beat frequency oscillator (BFO) type. Both of these operate from the magnetic field of a search coil using low frequencies of 1 kHz or less.

The T/R types use a transmitter coil and a receiver coil that are located in the search head. The transmitter coil, which operates as an antenna, provides a magnetic field which is affected by metallic objects near the search head. The perturbed field pattern is detected by the receiving coil and amplified.

The BFO type also uses a search coil in the head. The operating frequency of the search coil will change if a metallic object is near enough to the coil. This change is detected with a reference frequency. With some practice, various degrees of differentiation are possible between metal objects of different sizes as well as materials of different compositions.

The response of these units is generally too slow to detect metal fragments in a flowing process stream, where other techniques such as density measurement may be faster and more accurate, but also more expensive.

Metal detection in an actual process stream is an unusual concern. The more likely problem is detecting aggregates in a process stream. Depending on the size of the particulates this would be accomplished using one of the following: (1) mass flow measurements, (2) density measurement, or (3) mechanical filtering. The first two lend themselves to microcomputer control, which can facilitate rapid and accurate detection of small changes in the density or mass of the process stream.

Metal detectors are useful in locating underground piping or cabling such as gas and oil pipelines or electric utility and telephone cables. Some high-sensitivity detectors can be used to locate pipelines 10 to 20 ft (3 to 6 m) or more below ground level. Metal detectors are also used by geologists seeking metal and mineral deposits.

NOISE SENSORS

Acoustic noise is composed of oscillations in an elastic medium which occur in a frequency range in which the human ear is sensitive. This range is normally 20 to 20,000 Hz. The elastic medium is often air, and sometimes a liquid such as water.

Sound waves can appear as pressure, particle-velocity, or particle-displacement waves. Because of the limited compressibility of water, only pressure waves are usually measured in this medium. The pressure variations are usually small. At the normal threshold level of human audibility, the root-mean-square (rms) value of pressure is about 20 $\mu N/m^2$. Atmospheric pressure is about 200,000 N/m^2. The sensing device must not respond to this steady pressure.

The extremely wide range of sound pressures which need to be measured requires a scale of logarithmic units to describe sound pressure levels. The measured pressure is compared with the pressure at the threshold of hearing, 20 $\mu N/m^2$, using the following:

$$P_s = 20 \log_{10}(P_m/P_r) \text{ dB}$$

where

P_s = sound pressure level,
P_m = measured pressure,
P_r = reference pressure.

The conversion of pressure variations into an electrical output can be accomplished by several transduction techniques

1. inductive or reluctive
2. dynamic
3. electrostatic or capacitive
4. piezoelectric
5. resistive

OTHER PHYSICAL MEASUREMENTS 219

Some inductive sensors use the variation of reluctance in a magnetic circuit. These units use a suspended iron armature which is allowed to vibrate and change the air gap in the magnetic circuit. A pickup coil around an element of the magnetic circuit will have a waveform representative of the waveform of the sound. These inductive or electromagnetic microphones can be used for low-quality, high-level applications. They generally are not accurate enough for many industrial measurements.

Ribbon microphones use a thin, lightweight metal ribbon which is suspended in a magnetic field (Fig. 9-7a). Sound waves striking the ribbon cause it to move and cut through the magnetic lines of force, thus generating a voltage. The ribbon can be made very thin, and its light weight allows it to follow the motion of the air particles. Generally the entire area of the ribbon is exposed to the sound waves, so its velocity can be very close to the velocity of the air particles.

Dynamic or moving-coil microphones use a lightweight diaphragm that includes a coil of wire which moves within a magnetic field (Fig. 9-7b). The pressure of the sound waves as they strike the diaphragm causes the coil to move and a voltage is produced at the coil terminals. These units are classified as pressure actuated, since the diaphragm and coil move in response to sound wave pressure.

Capacitor or condenser microphones operate on the electrical charge held in a capacitor. As the capacitance is changed, the electrical potential between the condenser plates changes. If the capacitor consists of two plates, with one movable, the movement of one plate will change the capacitance and the potential between the plates will change in response.

The change in air pressure due to the sound waves operates the capacitor in the microphone. The initial potential may be supplied from a separate source through a high resistance or the electrical charge may be provided by making either the backplate or the diaphragm from an electret material that has a permanent electrical charge.

The carrier type of capacitor microphone is structurally similar, but no po-

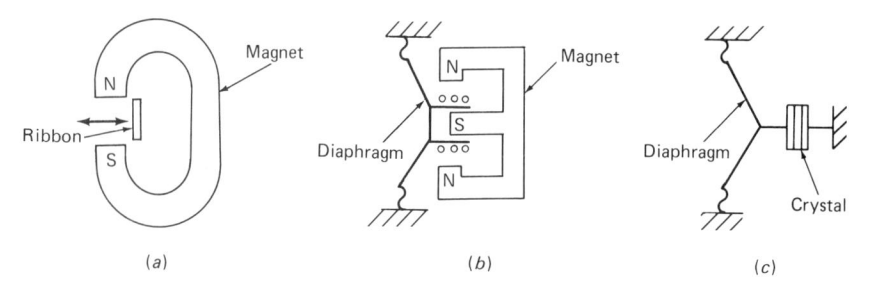

Figure 9-7. Microphone construction. a. Ribbon type. b. Moving coil. c. Piezoelectric.

larizing potential is used. The capacitance of the microphone is part of an oscillator. A change in capacitance causes a change in the oscillator frequency. This is converted into the desired output. This design allows operation at steady pressures down to zero frequency, which is not possible with the standard capacitor microphone.

The piezoelectric microphone uses a phenomenon characteristic of certain crystalline materials. A deformation of the crystal causes an electrical potential on the surfaces of the crystal. The magnitude of this potential is a function of the force on the crystal. Figure 9-7c shows a typical unit. A diaphragm receives the sound waves and imparts the force to the crystal.

The crystal element is normally an assembly of thin slices which is known as a bimorph. Piezoelectric microphones may use crystal elements made of Rochelle salt or ammonium dihydrogen phosphate. Ceramic materials have also been developed especially for this use. These microphones may be called electrostrictive rather than piezoelectric, but they operate in the same way.

Another common microphone is the resistive devices such as those used in carbon telephone handsets. Its characteristics make it suitable for voice communication but it is unsuitable for most other uses.

Sometimes microphones are classified according to the type of response they exhibit, which may be a function of the velocity, displacement, or pressure of the sound waves.

The ribbon microphone normally responds to velocity. The moving ribbon is light and flexible and easily moved. No mechanical damping is used and the motion is very close to the air particle motion so the output is proportional to velocity.

If the sound waves are allowed to pass through a pipe before they reach the ribbon this type of microphone can be made to act as a pressure microphone. The pipe can be damped such that it produces an acoustical resistance, which can be used to control such microphone characteristics as frequency sensitivity. The velocity type of ribbon microphone does not offer this feature. Its damping characteristics depend directly on the construction and it is generally not sensitive to large changes in pressure.

Most microphones with elastic diaphragms respond to pressure. This includes (1) the velocity of the coil motion in the dynamic microphone, (2) the diaphragm displacement in the condenser microphone, and (3) the crystal deformation in the piezoelectric microphone. These all respond in proportion to the applied force, which is the instantaneous sound pressure. However, the microphone structure must be designed so that the elastic deformation is linear with the proper damping. This is usually done by allowing air leakages through small tubes which add the correct acoustical resistance from the interior of the microphone to the outside air.

Pressure sensing microphones should measure pressure at a point. Thus, a smaller sensing area is desirable, but the diameter of the microphone diaphragm must be greater than the minimum wavelength of the sound.

The direction from which sound waves reach a microphone usually affect the output. A series of waves striking the diaphragm at a high angle of incidence will have a much smaller net effect than that produced by the same series of waves striking normal to the surface.

Pressure-gradient microphones operate on the difference in pressure between two points. They exhibit very specific directional characteristics and are useful in special-purpose applications.

When extremely high directional characteristics are required, other types such as the wave microphones can be used. These include the line microphone and the parabolic reflector microphone. The line microphone uses an array of pickup tubes of varying length. The sound is collected by the tubes and added vectorially as it reaches the microphone. These units have a high directivity but there is some distortion of frequency response due to the tube configuration. This is shown in Figure 9-8 along with the concept of the array.

Other sound-measuring devices are available which depend on the energy of the sound waves. These include (1) the hot-wire microphone, (2) the Lindval glow microphone, and (3) the Rayleigh disk. They are generally applicable to special-purpose applications although the Rayleigh disk is useful for calibration.

Ultrasonic microphones respond only to frequencies above the human hearing range, from about 25,000 to 45,000 Hz. They are useful for remote-control guidance devices applications and gas-leak detection. They generally use piezoelectric cyrstals which are resonant at the ultrasonic frequency. Typical microphone characteristics are listed in Table 9.4.

Microphones can also be classified as omnidirectional, unidirectional, or bidirectional. An omnidirectional microphone has a nearly uniform sensitivity to sound in all directions (Fig. 9-9a). This type of microphone would be used for measuring room or chamber sound levels. Microphones with a bidirectional response (Fig. 9-9b) are most often used for broadcasting and stage perfor-

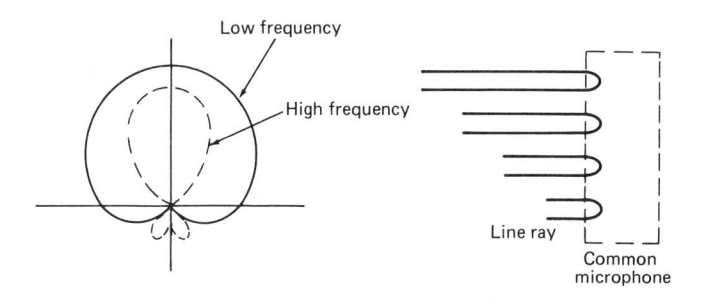

Figure 9-8. Line microphone directivity.

TABLE 9.4. Microphone characteristics.

Type	Range (Hz)	Output (dB[a])	Application	Features
Carbon	300–4000	−40	Voice	Low cost, voice quality
Ribbon	20–15,000	−80	Recording	Wide range, sensitive to pressure and velocity
Capacitor	12–15,000	−50	Sound-level measurements	Stable, wide range, flat response
Carrier capacitor	0.1–20,000	Variable	Special sound-level measurement	Very wide range
Piezoelectric	30–12,000	−60	Public address systems, recording	Good range, may be affected by temperature
Wave	80–8000	−80	Broadcasting	Highly directional

[a] 0 dB = 1 V/dyne/cm^2.

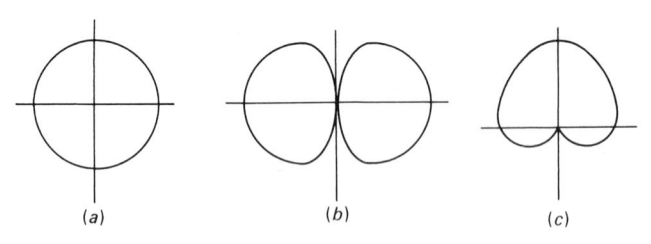

Figure 9-9. Microphone directivity patterns. *a*. Omnidirectional. *b*. Bidirectional. *c*. Cardioidal.

mances. This type of response can be provided by a ribbon microphone if both sides of the ribbon are exposed to the sound. With only one side exposed, the response pattern becomes cardioidal and takes the form shown in Figure 9-9c.

The frequency range of a microphone depends on both its operating principle and its construction. Most modern microphones exhibit a relatively flat response over the usual human hearing range. Dynamic microphones and diaphragm-type crystal microphones require good design and construction principles, including the proper damping, to provide a uniform response with no noticeable resonance peak. Condenser and ribbon microphones have a much smaller moving mass and fewer mechanical constraints.

The maximum amplitude for a particular microphone is a function of its construction. Mechanical damage can be caused by pressures that are too high. Most microphones can operate at the same maximum levels as the human ear for short periods of time. This is about 140 dB in terms of the sound pressure level. Special devices are also available which can be used continuously at much higher levels.

There are special microphones for use under water. These microphone units are designed in such a way that they respond to the vibratory sound components of pressure but not to the pressure due to the depth of immersion. These units are called hydrophones.

ENVIRONMENTAL CONSIDERATIONS

In some applications microphones may be exposed to temperature and humidity, mechanical shock and magnetic fields. These factors can affect the sensitivity and frequency response, either as a temporary or a permanent condition. Mechanical shock can affect delicate suspensions or cause crystals to become cracked. Dust can affect the small equalizing tubes and decrease the frequency response. Strong magnetic fields can weaken the permanent magnets in dynamic microphones and affect their sensitivity. Weaker magnetic fields can cause noise pickup by either the microphone or its cable. Ribbon microphones are especially sensitive to noise pickup from alternating currents. Crystal microphones are less sensitive to temperature effects, but strong electric fields can cause noise. Condenser microphones are stable under most conditions, but moisture and humidity can affect their response. Environmental wind can produce noise in all microphones. It can be reduced with windscreens that are installed on the microphone.

CALIBRATION

The proper calibration procedures can be used to detect, and often correct, many of the environmental changes that occur over a period of time. Microphones intended for measurement are usually supplied with detailed calibration data. The electronic equipment used with the microphone becomes a part of the calibrated system. Portable calibration devices are available for most types of microphones. They are usually small sound sources which are used on individual microphones.

Microphone sensitivities are usually expressed in terms of decibels referred to 1 V/dyne/cm^2. One dyne per square centimeter is about 75 dB. The sensitivity of microphones can also be given in volts per unit of pressure. A typical condenser microphone would exhibit about 10 mV for a sound pressure input of 1 N/m, which is equal to −40 dB.

STRESS AND STRAIN MEASUREMENT

Stress and strain are generated in many parts and mechanical systems by weight, temperature, pressure, vibration, or displacement forces. The most common method of making these measurements is by the use of strain gauges. Other techniques include stress coating and photoelastics.

If a load or stress is placed on an object, the object may expand or contract or be subjected to shear. When a grid of wire or foil with the proper resistive characteristics is bonded to the object, it will tend to stretch or be compressed as the surface to which it is connected expands or contracts.

The metallic resistance strain gauge operates on the principle that, as a conductor is subjected to a tensile or compressive strain, it will also experience a proportional change in resistance. The magnitude of this resistance change follows a change in the magnitude of the applied strain, which is defined as

$$\text{Strain} = \frac{\text{Change in length}}{\text{Original length}}.$$

An important strain gauge property is the constant known as the gauge factor (GF). This constant can range from 2 to 6 for the most common strain gauge alloys. The gauge factor is the ratio of the change that occurs in the total resistance as related to the change of length in the conductor with respect to its total length as indicated below:

$$\text{GF} = \frac{\Delta R/R}{\Delta L/L}.$$

METALLIC STRAIN GAUGES

Resistive metal wire strain gauges may be bonded, surface transferable, or welded. The unbonded strain gauge, which consists of a wire stretched between posts and unsupported between these points, is used in some pressure transducers but generally not for strain measurements.

The bondable type of strain gauge is usually bonded with cement to the measured surface. It is the most common type of strain gauge.

A flat wire grid may be used which consists of a thin wire or filament arranged in a winding or pattern and cemented to a base. The lead wires are soldered or welded to the ends of the filament to provide the electrical connections. The length of the grid is the gauge length. A cover plate may be cemented over the wire grid.

The choice of the filament material is usually based on five factors:

1. The resistivity of the material should be high to allow winding within the smallest possible area.

2. The temperature coefficient of resistance should be low to minimize temperature errors.
3. The material should provide as high a gauge factor as possible so that its resistance change with strain is large.
4. The material should generate the lowest possible thermoelectric potential at the junctions with the leads.
5. The filament should have a high mechanical strength to allow high stresses to be applied to the filament.

Gauges with multiple grids called rosette gauges can be used for the simultaneous measurement of strain in different directions. These gauges use a sandwich construction with the grids laid over each other and insulated by cement.

Weldable strain gauges are mounted using a small flat or tubular envelope made preferably of the same metal as the surface to be measured. The base of the envelope is slightly larger than the main body of the gauge for welding to the surface. Since the base is thin and the edge is narrow, microwelding techniques are normally used for mounting weldable gauges.

These gauges are used on the inside and outside surfaces of tanks containing liquids. Strain gauges for strain measurements in concrete can be encapsulated in a waterproof container to protect the filament during installation when pouring the concrete.

Metal foil gauges use photoetching techniques similar to those used for printed circuit boards. Foil gauges allow better utilization of a given area and can be made smaller than wire grid gauges. They can also be used to measure larger strains than wire types and are not damaged as easily. For many applications, foil gauges are superior in all of the most essential areas.

The foil grid consists of a 3 to 8 μm layer of foil, some of which is removed by etching to the desired grid shape. The grid can be cemented to a base. The use of the etched-foil technique allows a number of different designs including dual- and triple-element rosettes as well as biaxial gauges.

Some biaxial gauges have two elements spaced 90° from each other and at 45° angles from an identifiable center line. This spacing allows torsional strain measurement under an applied torque. Triaxial rosettes with elements spaced at 60° from each other are also available. Spiral foil gauges are available for measuring tangential strain in a diaphragm. The foil pattern is bifilar to facilitate its construction and to cancel inductive effects. Full bridge rosettes may use opposing biaxial combinations for structural applications, or an in-line configuration for bending beam applications.

A stress-strain gauge will use two uniaxial strain sensing elements oriented at 90° to each other. The elements can have a common connection so that either of them may be observed independently for conventional strain measurement or the series combination of the two elements can be used to produce an output proportional to stress along the principle axis is shown in Figure 9-10. This last

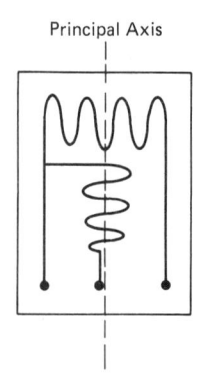

Figure 9-10. Two-element stress-strain gauge.

measurement is achieved by making the ratio between the resistances of one element to the other element equal to Poisson's ratio of the material to which the gauge combination is to be applied. If the stress-strain gauge is to be applied to aluminum, with Poisson's ratio 0.33, the resistances of the elements are 115 and 345 ohms. Strain in either the principal axis or the transverse axis is measured in the usual manner using the gauge factors for each element. The stress is determined by using a manufacturer-supplied stress gauge factor.

Thin-film techniques are also used to manufacture strain gauges. Here a thin film is applied directly to the measured surface.

Thin-film gauges can be made smaller than equivalent metal strain gauges. The surface is first coated with an insulating substrate upon which the gauge or rosette is formed by evaporative or bombardment methods. Applications of thin-film metal straining gauges have been mainly to the diaphragms of pressure transducers, but this technique can be extended to other sensing elements.

Flame-sprayed strain gauges are used mostly for structures which are exposed to hostile environments. In these applications they have been used as an alternative to welded strain gauges although the installation techniques are more costly.

Flame-sprayed gauges are thinner than other strain gauges and this can be an advantage in some applications such as those surfaces which may be subject to aerodynamic heating effects.

A strain-sensitive metal grid is applied to an insulating ceramic substrate which has been formed on the measured surface. The substrate is formed using a process in which the end of a ceramic rod is heated and the molten particles projected onto the surface. Substrates 0.05 to 0.1 mm thick have been used on some engine applications.

A surface transferable metal gauge is then applied to the substrate and permanently bonded to the surface with a flame-sprayed ceramic coating. Temperatures up to 2200°F may be used for these flame-sprayed strain sensors.

SEMICONDUCTOR STRAIN GAUGES

Semiconductor strain gauges make use of the piezoresistive effect, which is much larger in semiconductors than in conductors. The gauge factors of semiconductor strain gauges range between 50 and 200, while those of metal strain gauges are usually no greater than 6 and often are as low as 2. The disadvantages of semiconductor gauges are

1. Operating temperature range is limited.
2. They require more temperature compensation.
3. Strain ranges are usually limited.
4. They tend to be more difficult to apply to measured surfaces.

The gauge factor is dependent on the resistance change due to dimensional change as well as the change of resistivity with strain. It also tends to be nonlinear with the applied strain. This nonlinearity is usually minimized by using a strain gauge bridge circuit.

One bridge technique is to use a pair of active gauges, matched to each other, as adjacent bridge arms. Others use the cancellation of nonlinearity by an opposite nonlinearity in a bridge circuit or use a four-active-arm bridge with two gauges in tension and two in compression with gauges that are matched, or by shunting a single active arm.

Another method requires prestressing the gauge in tension during mounting to shift its operation to a more linear portion. The amount of doping can also be increased to obtain better linearity with a reduction in gauge factor. The linearity of highly doped gauges approaches that of most metal gauges.

Semiconductor gauges are more sensitive to temperature changes than metal gauges. The temperature coefficient of resistance of the semiconductors used is 60 to 100 times greater than that of Constantan. The variation of gauge factor with temperature is 3 to 5 times greater, and the Seebeck coefficient due to the thermoelectric potential generated at the connections can be 10 to 20 times as large. The differential thermal expansion between the semiconductor and the measured surface may also be greater. The linear expansion coefficient of a semiconductor gauge is about half that of a metal gauge.

Temperature compensation for semiconductor gauges can be achieved by a number of techniques:

1. A dummy gauge can be connected to one bridge arm.
2. Thermocouples can be used to provide an opposite voltage.

3. Resistors or thermistors with controlled thermal resistance changes can be connected in series or parallel with the active semiconductor gauge bridge arms or with one or both of the excitation leads.
4. A *p*-type semiconductor with a positive gauge factor and temperature coefficient can be connected in an active bridge arm when using an active *n*-type gauge.
5. Gauges can be selected such that the thermal expansion and temperature coefficient of resistance tend to balance each other.

Typical semiconductor strain gauge configurations are shown in Figure 9-11. The gauge may be surface transferable or encapsulated. Gauge lengths vary from 0.01 to 0.25 in. Lead materials are usually gold, copper, or silver wire.

Silicon is commonly used for the semiconductor material. Some designs fuse the dopant directly into portions of silicon blocks or diaphragms. These sensing elements with integral diffused strain sensors have been used in pressure and force transducers.

STRAIN GAUGE CHARACTERISTICS

Electrical characteristics include the gauge resistance, power rating, or maximum current, which depends on the heat sink to the measured surface and the bond. The insulation resistance is an important parameter for weldage gauges, since no insulation is provided by bonding cement. The gauge resistance values are typically 60, 120, 240, 350, 500, and 1000 ohms for metallic gauges, and for semiconductor types 5000 and 10,000 ohms are common.

An installed strain gauge senses the strain in a plane which is separated from

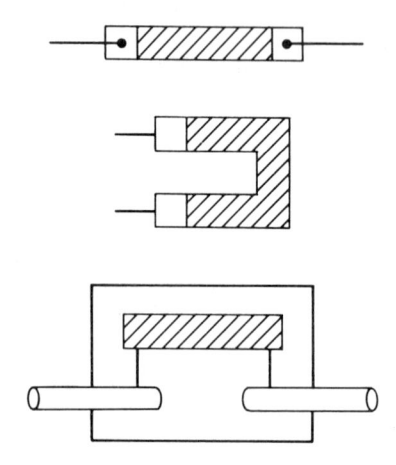

Figure 9-11. Semiconductor strain gauge configurations.

the measured surface by cement and the mounting surface. If axial strains are being sensed, the strain path lies on the surface and it is transmitted through the cement and mounting surface to the gauge.

If the specimen to be measured is under a bending stress, the strain at a surface is defined by the distance of the surface from the neutral axis and the radius of curvature of the axis. As shown in Figure 9-12, the gauge then lies farther from the neutral axis than the measured surface. The gauge will now sense a strain larger than that occurring on the specimen. This is called the offset error.

The offset error can be minimized by placing the gauge as close to the specimen as possible. The center plane of a wire grid gauge can be approximately 0.005 in. above the specimen surface while the center plane of a foil gauge is approximately 0.002 in. above. Thus a foil gauge will have about half of the offset error of a wire gauge. In many cases the offset error can be eliminated with the proper calibration.

The major ambient temperature performance characteristic of all strain gauges is the gauge factor. It is usually specified as a nominal value with tolerances.

The complete specifications should cover the strain range for which the linearity, hysteresis, and creep tolerances apply, as well as the allowable overload ac strain limit that will not cause output errors. Linearity may apply only to a specified portion of the range. Hysteresis and creep are usually expressed in units of strain. Drift refers to the changes in gauge resistance at a constant extreme temperature with no applied strain.

Most strain gauge applications require a bridge configuration; thus the electrical characteristics of the strain gauge element and the degree of matching required from the other elements as well as the internal geometry of the combination become important. Active/dummy combinations are commonly used for both types of elements.

Dimensions tend to be more critical for strain gauges than for many other types of transducers. The gauge length and width tolerances must be commensurate with the magnitude of the dimensions. If the gauge is furnished with leads, the diameter or thickness of the leads, lead material insulation, and length and spacing between leads can be important. If the gauge is bonded to a carrier

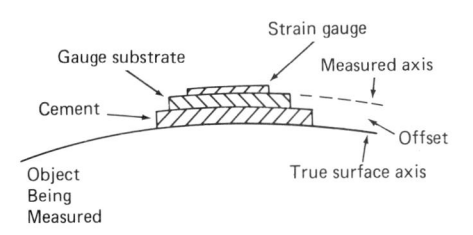

Figure 9-12. Offset error.

230 TRANSDUCERS FOR AUTOMATION

or is encapsulated, the base length, width, thickness, and material must be known.

In gauges that are surface transferable the size and material of the strippable carrier become important. The material as well as the grid wire diameter or fill thickness are important considerations.

The recommended methods for bonding or other attachment of the strain gauge to the measured surface or its embedding in specific materials should be known. This information can include (1) type of cement, (2) cure time, (3) temperature and pressure, (4) material compatibilities, (5) insulation resistance and changes at temperature extremes, (6) resistance to humidity and moisture proofing.

The measurement of strain using a resistance strain gauge requires the accurate measurement of a small change in resistance. The most common circuit used to achieve this is the Wheatstone bridge.

The voltage across the output is zero for a balanced bridge and has a predictable and measurable amplitude when one arm of the bridge changes. Since the resistance change ratio is related to the applied strain, the output voltage is also a function of the strain. The relationship between voltage and strain can be kept linear with the proper choice of operational limits.

The output voltage is usually a few millivolts, which makes it necessary to use high-sensitivity amplifiers in order to increase the signal level. The signal/noise ratio must also be controlled to produce a reliable signal output.

A typical circuit is shown in Figure 9-13 for a dc excitation system. Only one active gauge is shown, but two, three, or four gauges may be active; unstrained gauges known as dummy gauges can also be used for compensation.

Gauges may be used in half-bridge or full-bridge configurations. Figure 9-14 illustrates the use of two active gauge circuits, and Figure 9-15 illustrates the use of four active gauge circuits. These circuit configurations are used mainly for static strain measurements.

There are a number of techniques which can be used to calibrate strain gauges and strain gauge systems. The most basic method is to place known loads on

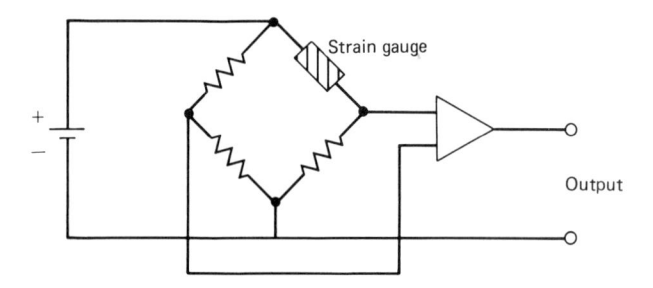

Figure 9-13. Direct-current strain gauge system.

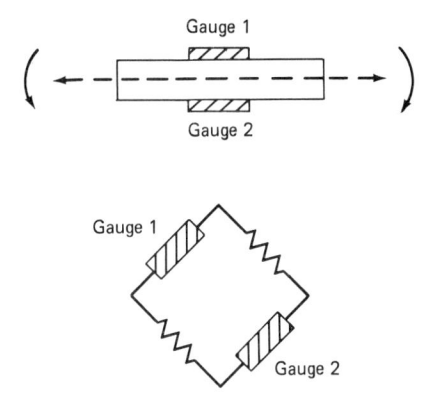

Figure 9-14. Double-active bridge configuration for an axial load with bending compensation.

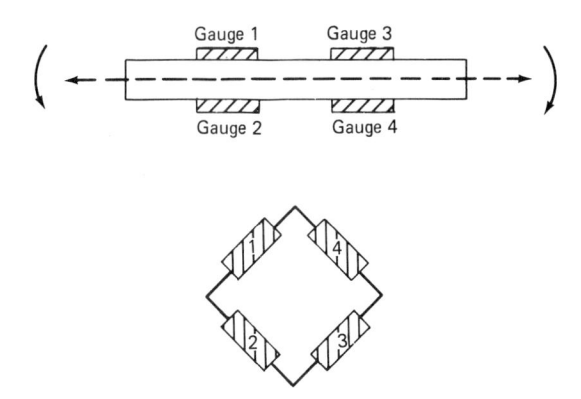

Figure 9-15. Quad active bridge configuration for an axial load with bending compensation.

the gauges and measure the output. In many applications this method cannot be used, since the system cannot be exercised to the calibration values. If one calibrates the gauges to eliminate all the errors that can exist, a computer model may be required. Errors can occur as a result of the following:

1. the temperature coefficient of the materials
2. sensitivity changes with temperature variations and the methods of correction
3. calibration errors caused by series or shunt resistances
4. resistance of transmission lines and the configuration

In many applications when the half-bridge configuration is used, the following expression provides a calibration value:

$$R_c = \frac{R_g}{\epsilon K} - R_g.$$

If the signal leads are long enough to have a significant resistance this equation becomes

$$R_c = \frac{R_t}{K} - R_t,$$

where

R_c = calibration resistance (ohms),
R_t = total resistance of gauge and leads (ohms),
R_g = gauge resistance,
K = gauge factor,
ϵ = strain value (μin./in.).

VISUAL TECHNIQUES

Photoelasticity is a visual technique which can be used to indicate strain and stresses in parts and structures. If a photoelastic material is subjected to forces and viewed under polarized light, the resulting strains appear as fringe patterns. The patterns show the overall strain distribution, and quantitative measurements can be made of the strain directions and magnitudes. The three main photoelastic techniques are (1) two-dimensional model analysis, (2) three-dimensional model analysis, and (3) photoelastic coating analysis.

The photoelastic coating method is the most widely used since it allows the use of photoelasticity on actual parts of any shape under actual or simulated service conditions. Two-dimensional analysis is used to determine stress concentration factors and three-dimensional analysis is used to provide a complete stress analysis both inside and outside the part.

Brittle coatings are lacquers which are applied by spraying on parts or structures. After drying, the part is stressed and the coating will exhibit fine cracks, showing the location of maximum strain and the direction of the principal strains.

This method is not suitable for detailed quantitative analysis like photoelasticity. Brittle coatings are generally used to determine in advance the exact areas for strain gauge location and orientation.

Moiré fringe analysis may be used to determine the components of displacement or strain. First a grid of equidistant lines is deposited on the part to be stressed. Any deformation of the part also affects the grid applied to it. The deformed grid is then superimposed and compared to an undeformed grid. This superposition produces an optical effect known as moiré fringes.

This technique is generally applied to the measurement of strains that cannot be determined easily, accurately, or economically by other techniques such as high-temperature strain measurements and the measurement of large elastic and plastic strains.

FORCE AND TORQUE TRANSDUCERS

Most force and torque transducers use a separate sensing element which converts the measured quantity into a small mechanical displacement. This is usually done via a deformation of an elastic element. An elastic deformation used to sense force in this way is a strain or deflection. A maximum of each of these will occur at some location in the sensing element but not necessarily at the same location.

This maximum of either strain or deflection is applied to the force or torque element. The element must be of a homogeneous material and usually manufactured to close tolerances. A number of different types of steel are commonly used as the element materials.

Beams are among the simplest types of force sensing elements. The maximum deflection of a beam always occurs at the point of applied force. The maximum deflection of a cantilever beam occurs at its free end and the point of maximum strain is at the fixed end.

Cantilever beams of constant strength have a triangular or parabolic tapered shape which is narrowest at the point of force application. In these constant-strength beams the strain is constant along the bottom of the beam. In the simply supported beam of the point of maximum strain is at the point of force application.

Column force sensing elements generally have the point of maximum deflection at their vertical center at one-half of the column height. Their characteristics depend mainly on the height to width ratio and on the wall thickness for hollow cylinders. Compression and tension in a column may be sensed as strain or as changes in magnetic characteristics or natural vibration frequency, but they are rarely used as displacement changes.

Proving rings and the square, flat type of proving rings or frames are often used as force sensing elements. These are shown in Figure 9-16. The maximum deflection as well as the maximum strain occur at the point of force application. The strain has almost an equal magnitude at points 90° in either direction from

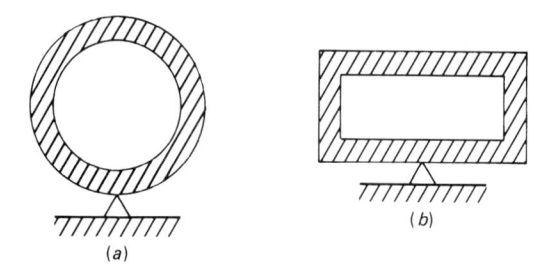

Figure 9-16. Proving-ring force elements. *a*. Standard ring. *b*. Flat or frame type.

the point of force application, and the strain sensing is generally more convenient at these points.

Diaphragms can also be used as force sensing elements. Maximum strain and deflection occurs at the center of the diaphragm, where the force is always applied.

PIEZOELECTRIC FORCE TRANSDUCERS

Piezoelectric elements can be used for force sensing. Although piezoelectric force transducers can be used for a measurement of quasistatic response by using charge amplifiers, they are mostly used for dynamic force measurement. Column-type sensing elements are normally used with a ceramic or quartz transduction element located at the center of the cylindrical column.

In some units a pair of disk-shaped piezoelectric crystals is separated by a thin electrode. The electrode is insulated from the case and connected to a coaxial cable. The crystal assembly is between two thick metal disks which are threaded for mounting.

Another unit is the washer type, which uses quartz segments as shown in Figure 9-17. The sensing element takes the form of an annular column with a sandwich construction. The central hole may be used with a bolt or stud. The force washer unit is often used with threaded fastener mountings of machinery and other equipment.

The piezoelectric transducer responds to compression forces only. It can be preloaded so that a force is continuously exerted producing a static level. If a bidirectional dynamic force is then applied, the force will provide an output which is a function of the alternate tension and compression. Piezoelectric force transducers do not require an excitation. They have a very high output impedance and voltage amplifiers, emitter followers, and charge amplifiers are used with these transducers. This equipment has been discussed in the chapters dealing with piezoelectric acceleration and pressure transducers.

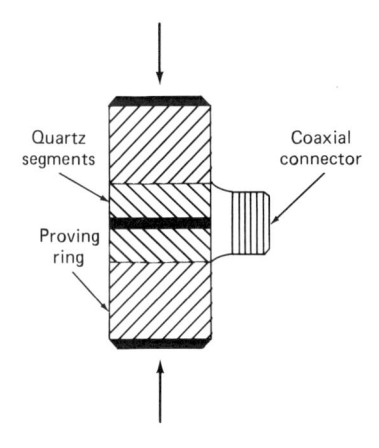

Figure 9-17. Piezoelectric force transducer.

RELUCTIVE FORCE TRANSDUCERS

Reluctive force transducers use the deflection of a mechanical sensing element. Some of these transducers use differential transformers.

The deflection of a proving ring or diaphragm causes a relative motion between the core and a coil. This produces a change in the amplitude and phase of an ac voltage in the secondary windings.

The coil assembly is usually encapsulated and shielded. The excitation voltage ranges between 2 and 10 V and the frequency between 50 Hz and 10 kHz. The sensitivity improves with increasing excitation frequency. With the secondary windings connected in the most common configuration, the output voltage increases equally from the core center null point for sensing deflection to compression or tension, and the output phase changes from 90° toward zero degrees in one direction and toward 180° in the other direction. A phase-sensitive detector can be used to improve the output accuracy.

CAPACITIVE FORCE TRANSDUCERS

Capacitive force transducers detect force by a change of capacitance. These transducers tend to require more complex circuitry than strain gauge or even reluctive transducers. They are used mainly for dynamic force measurements where a wide frequency response is needed or in specialized applications, such as measurements in low-pressure and high-temperature environments.

Grounded beams and diaphragms are used as rotors in many capacitive transducers. The capacitance changes may be converted into changes of a dc output voltage by signal conditioning. Transducer operation tends to be more effective at the higher excitation frequencies since large changes in reactance will occur.

The variable capacitor represented by the transducer can be one leg of an ac bridge or it can be (1) a shunting capacitor across an amplifier input, (2) a feedback element in an ac amplifier, or (3) an element of a tuned *LC* circuit. In the typical transducer shown in Figure 9-18, the capacitive element may be used with a coil to form a tuned tank circuit. One end of the moving capacitor (the rotor) is grounded. The rotor forms one end of the hollow cylinder force sensing element. The transducer stator forms the other end.

A voltage in the megahertz range is coupled to the tank circuit through a matching coil. Changes in capacitance are then detected as variations in the impedance of the matching coil due to the changes required to tune the tank circuit.

Vibrating-wire force transducers use a wire between a pair of permanent magnets. The wire is stretched between two points of a deflecting force sensing element. The wire is forced to vibrate at its resonant frequency in a feedback oscillator and the oscillator frequency changes with the wire tension.

This type of force transducer is normally used with a frequency-modulated output, primarily in countries other than the United States. If a second wire which is not affected by the sensing element deflection is allowed to oscillate at the same frequency as the sensing element wire when no force is applied, the beat frequency, which is the frequency difference between the two frequencies of oscillation, will be a function of the tension in the sensing wire alone.

TORQUE TRANSDUCERS

In-line torque sensors for applications with rotating shafts use special sensing shafts which are inserted between the mechanical power source and the load. As a torque is applied to one end while the other end is fixed, a line on the shaft surface which is originally parallel to the axis of rotation will become a helix or a part of a helix. This same twisting action occurs during normal use

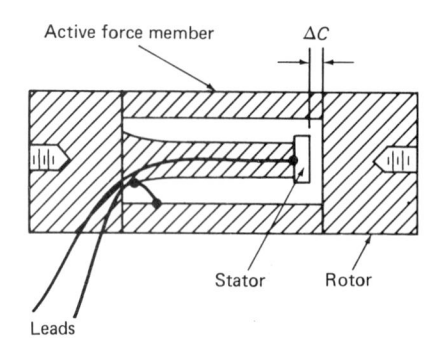

Figure 9-18. Capacitive force transducer.

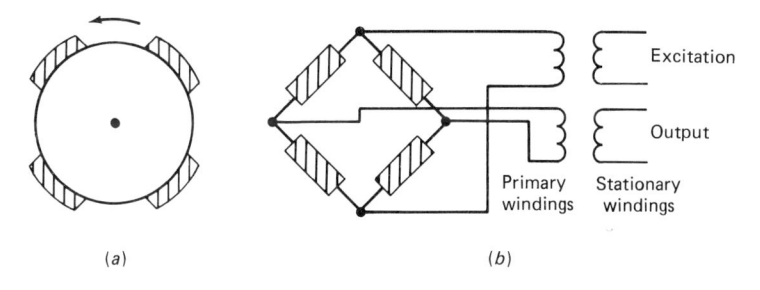

Figure 9-19. Strain gauge bridge mounted on rotating shaft. *a.* Configuration. *b.* Connections.

of the shaft at a much lower magnitude. It may be sensed as surface strain as illustrated in Figure 9-19. The connections may be through slip rings and brushes or transformers.

Torque sensing units for shaft torque measurements use the surface strains or stress patterns of the shaft much more often than deflection. Torque sensing units can be used to measure torque regardless of the direction of shaft rotation.

Force transducers can respond equally or unequally to tension and compression and some designs can be used for one force direction only. More strain can be obtained by machining a section of a diameter of the shaft and sensing the strain on the surface of this section only.

A square notch can be used for strain gauge mounting with electrical connections in the low-stress area near the corners. A notch with linked rods is used with special transduction elements. One type contains four rods of square cross section. This type of torque sensing element provides some performance improvements over other types when used with strain gauges.

Encoders can be used as torque transducers. These torque measuring units use the disks of incremental digital angular displacement transducers. Two disks are mechanically attached to the sensing shaft so that as the shaft deflects due to a torque a relative angular displacement occurs between the disks.

When toothed ferromagnetic disks are used, the phase difference between two electromagnetic or reluctive output reading devices will be proportional to the shaft deflection. This same technique can be used with optically coded disks using alternating transparent and opaque portions with photoelectric reading devices.

Another photoelectric torque sensing method uses a single light beam through two disks. The disks are first aligned such that the combination of the opaque and transparent sectors through which the light beam must pass produces a null output at the detector. When a torque is applied to the shaft, the disks will pass either more or less light, depending on the null point. The output of the detector will be a function of the applied torque.

238 TRANSDUCERS FOR AUTOMATION

The critical design parameters of force and torque sensing elements include

1. Material.
2. Relative size and shape.
3. Modulus of elasticity.
4. Dynamic response.
5. Sensitivity in terms of strain and deflection.
6. Effects of transducer loading on the measured system.

First decide if a universal tension or compression, tension, or compression force transducer or a shaft torque or reaction torque transducer is to be used. Then select the required measurement principle. The following characteristics should also be considered in the selection:

1. Case.
2. Mounting dimensions and tolerances.
3. Torque and force connecting provisions.
4. Electrical connections.
5. Optional devices.
6. Operating environment.

Slip rings can wear and become dirty, which can result in erratic output. Transformer coupling can be used with ac-excited strain gauges or with four coils as shown in Figure 9-20 to detect changes in the permeability of the material. The permeability increases under tensile stress and decreases under compressive stress. A variation of this technique consists of many primary and secondary windings arranged in a ring around the shaft. This reduces errors due to residual magnetism and integrates the stress changes to produce a more representative output.

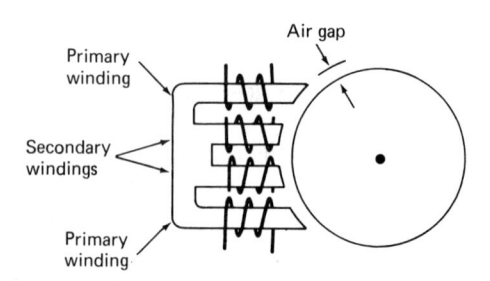

Figure 9-20. Measuring permeability with a transformer.

EXERCISES

1. Discuss the basic principle of turbidity measurement. What are some of the difficulties of measurement?
2. How can thermal and electrical conductivity be measured?
3. Discuss four methods that oxygen analyzers may use.
4. Describe two methods of determining pH with the aid of diagrams.
5. Discuss the basic principles used in flame sensors.
6. Describe how noise sensors can convert pressure variations to electrical signals.
7. What environmental considerations are important in noise measurements?
8. Why are leak detectors important in automation systems?
9. List and briefly describe 12 methods that can be used for leak detection.
10. What are some techniques of detecting metallic particles in a process stream? Which would be most easily employed with a microcomputer measuring system?
11. What is the importance of the gauge factor in a strain gauge?
12. Prove that the smaller the resistance of a strain gauge the greater the error a given lead resistance produces.
13. What are the problems involved in the event that all the bridge components of a strain gauge bridge are not simple resistors.
14. Discuss the advantages and disadvantages of semiconductor strain gauges.
15. If a circuit using a 350-ohm strain gauge, having a 4.0 gauge factor, is to be measured by means of a 50-ohm cable in a bridge circuit having matched gauge resistors, what current will flow for a 1800-μin./in. strain when 5 V is applied to the circuit?
16. What is the R_c value of a 120-ohm gauge with a 1.4-ohm lead resistance and a gauge factor of 4 for a 5000-μin./in. strain application?
17. Assuming a gauge resistance of 350 ohm, a lead resistance of 1.2 ohm, a gauge factor of 4.5, and a strain of 2000 μin./in., find the R_c value.
18. Describe the possible use of photoelasticity, brittle coatings, and moiré fringes in automated strain analysis using a microcomputer.
19. Show using a block diagram how a microcomputer could be used to reduce strain gauge errors during calibration.
20. Describe why the construction of the foil gauge grid and method of manufacturing the grid material should make it easier to control the properties of a foil than a wire. How does this result in a greater degree of reproducibility among gauges and better stability of the individual elements?
21. The vibration sensitivity of a strain gauge may be defined as the peak instantaneous change in output at a given vibration level. It is usually expressed as a percentage of full-scale output per "g" level over a given frequency range. It can also be specified as the total error in percentage of full-scale output for a given acceleration level. Define a test sequence for determining the vibration sensitivity of an unbonded strain gauge. Use a microcomputer for the control and sensitivity calculation. Write a flow chart for the microprocessor.
22. Draw a strain gauge bridge configuration for measuring the shear in a steel panel which compensates for bending forces. Describe how a column could be used to measure force using strain gauges to provide an output to a data acquisition system. How could the column be used to calibrate the force output?

240 TRANSDUCERS FOR AUTOMATION

23. Describe a self-contained capacitive force transducer. What are the interface requirements for connection to a microcomputer serial input?

24. A reluctive force transducer is to be used with a proving ring for measuring force. Describe with the aid of a diagram the physical and electrical connections for the system. The maximum sensitivity and accuracy are desired. Describe any special considerations.

25. Describe with the aid of diagrams two techniques of using photoelectric encoding for torque measurement. Which of these could provide digital information in the format required by a microcomputer? What additional circuitry or equipment would be required?

26. List the basic specifications required to specify a force or torque sensing element for an automation application.

10
AUTOMATION FUNDAMENTALS

COMPUTER-AIDED TECHNIQUES

The use of computers in manufacturing automation has been growing rapidly. Computer automation applications include inventory control, scheduling, machine monitoring, management information systems, and other information applications. These applications are primarily involved in transferring, interpreting, and tracking manufacturing data. The computers are used here primarily as data managers.

The major area of computer automation in manufacturing is that of controlling the physical manufacturing process. This area is typically classified as computer-aided manufacturing (CAM). This technology allows great economic benefits in improving manufacturing productivity but it also presents a number of technical challenges which are inherent to computer control of a physical process.

This technology provides an incentive for businesses to modernize plants and equipment. The average age of much industrial equipment is over 20 years and in mature industries such as steel, paper, and foundries, much of the equipment can be over 50 years old. By contrast, the average age of the industrial equipment of some newer competitors is about 10 to 14 years.

If an industry is to compete, it must use the most efficient tools available. The new computer technologies will encourage businesses to update their equipment. This also applies to equipment used in research and development, which is an important component of productivity growth, since over the long term only R&D can enable one to maintain technological leadership and maintain gains in productivity growth.

A factory automation program is a comprehensive contribution toward improving productivity. But a number of factors are involved in successful implementation of a system and management has the obligation to contribute in a number of ways. To take full advantage of such a program obsolete equipment that deters productivity growth must be replaced or updated.

The implementation of factory automation will not be easy for many segments of industry. Because of the complexity and size of many manufacturing operations any venture to improve the manufacturing process significantly requires a major capital investment and restructuring.

The use of integrated computer systems can have some serious impacts on corporations themselves and how they operate. Engineering and manufacturing organizations tend to merge together into a single design factory.

After the systems have been adopted, new challenges must also be overcome. Entire departments may have to be retrained to use the system.

The key to the future lies with computer-aided systems since the successful application of these technologies means building better products easier and faster. Many future products, in order to be more competitive, will rely on the use of advanced materials, particularly composites. These composite components may be made up of hundreds of layers of a material like graphite epoxy.

Computers will be needed to keep track of the shape, end points, and direction of the plys. Factory automation will provide the means to automate the equipment to produce these fabrications, decreasing costs while enhancing structural integrity.

Computer automation also allows improved tolerances, which can play a major role in reducing manufacturing costs. Complex aircraft structures are now fabricated without using a single shim. This was quite uncommon before the extensive use of computers in manufacturing. Many of these structures involve the drilling of over 20,000 holes using computer techniques. The rivets that hold the skins to stringers are applied without a single production snag. In this type of automation system the data are directed to the riveting machines to control their operation. Widebody commercial aircraft can require over 100,000 structural components. A description of these parts is provided in an integrated system that allows a complete flow of computerized data including completed drawings from sets of stored data that describe the drawings mathematically.

This type of system can also produce assembly and fabrication drawings, tooling information, and interior layouts. It may also provide aids that make such activities as testing and analysis easier. Correcting errors is always costly, but by using computer techniques to eliminate most of these expensive errors a great deal of time can be saved.

Computer automation also allows increased productivity as one of its major benefits. Productivity gains of up to five to one may be achieved in some applications. Higher gains are possible and gains of two to one are average.

Many control processes are candidates to undergo substantial changes and be significantly automated. Scanning devices and other sensors can be used to compare physical characteristics with a computerized data base and detect defects.

Computer techniques are also used in the scheduling and estimating processes. Statistics for component production can be used to provide the various manufacturing departments with data, and through the use of these computerized data the manufacturing operation is able to reduce overall costs significantly.

The advent of integrated computer systems also makes the increased use of robots more feasible. Integrated computer-aided manufacturing (ICAM) is primarily concerned with automated manufacturing, emphasizing robotics.

In a unified factory system, the entire manufacturing process is controlled from start to finish. Such capabilities are presently available in a few large manufacturing operations. But more and more plants are gaining these capabilities in an effort to modernize and streamline their operations.

DEVELOPING FACTORY SOFTWARE INTERFACES

The evolution of factory software encourages the creation of application packages that support the development of information systems for a variety of product requirements and changing manufacturing conditions. To help reduce product development time, emphasis must be placed on the quality of the transmitted data for computer-controlled manufacturing, product administration, and manufacturing planning.

This section will consider the implementation of application programs developed to support this process. The objectives and general format of the data structure used to describe the part geometry, documentation graphics, and bills of material for an automated release to manufacturing will be discussed.

INTERORGANIZATIONAL COMMUNICATIONS

The task of preparing manufacturing build data from design specifications has evolved from a basic batch programming capability to a highly interactive automated data base/data communications system. This evolution has occurred during the past twenty years. Figure 10-1 shows a typical high-level flow for the information systems involved in the design release and control process. Each task represents an information systems development organization and a set of application programs for providing data to administer product development. The software process can be considered as a part of the application programming strategy for processing the interface data.

Typical of organizational charters are the following:

1. A design automation group that develops programs for design capture, simulation, test generation, and the physical design.
2. A process automation group that develops programs to provide data for a computer-controlled manufacturing process.
3. Administrative systems are developed to support the product structure documentation requirements for high-level forecasting and manufacturing requirements planning.
4. A separate group is responsible for documenting the release methodology, obtaining design commitments for build data, and general coordination of this process. They help ensure that the developers work toward common business goals and that optimal decisions are made concerning the ownership of source data and the timing of entering manufacturing processes with derived data.

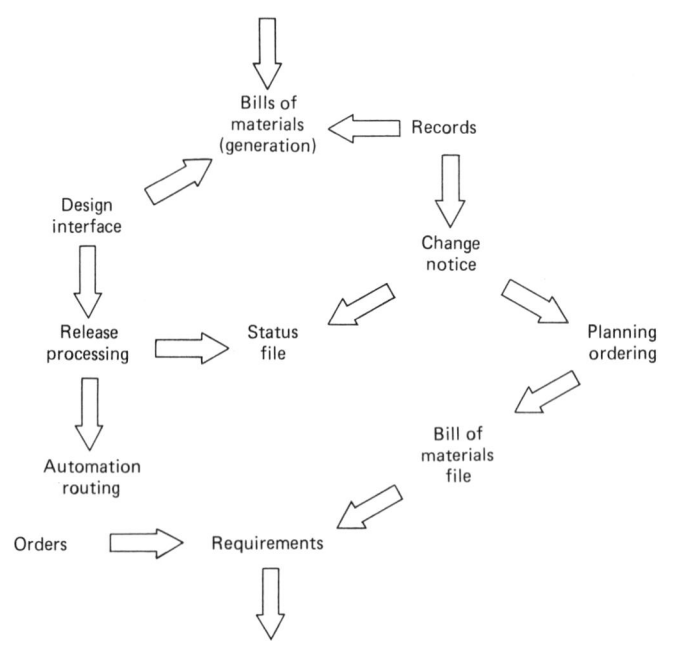

Figure 10-1. Factory information flow.

Refinements are continually taking place, driven by new products, processes, and quality and productivity improvements. The structure of the design manufacturing interface must support this changing environment and allow backwards compatibility for older versions of the data structure.

Conflicts are resolved by conducting work sessions with the input of the engineers, analysts, and programmers from the responsible organizations. The desired philosophy is to strive for decisions at the lowest level possible.

The systems might be implemented as shown in Figure 10-1. The initial timing of events begins with the bill of materials entering the system. Design completion results in a comparison and modification of the design structure and/or the administrative structure.

A library of release rules is administered which covers the release methodology. These rules direct the systems release action to produce the interface with the proper content.

An engineering change notice is sent to the group or groups that have been designated to build the package. This notifies them of the impending release interface. Upon the receipt of the interface the following take place: (1) manufacturing files are updated; (2) numerical control data are produced; (3) the change is implemented. If data errors are detected, the design groups are notified and the cycle is repeated until the data are correct. This cycle is usually known as the release process. The application software interface is then called the release interface.

THE DATA MODEL

In the ideal case the design systems have been developed to support a software model of the logical and physical structure. A real electro-mechanical product may require schematic representations, component placement information, and point-to-point wiring information to document the design and enable a manufacturing facility to build the product. During this development, the graphics may have been captured at gradually higher levels of the process based on an increasing graphics capability.

The original implementations may have ignored many variables. If the data were hard coded in programs as specific, positioned data for artwork or drafting generation, then little or no internal manipulation was allowed.

As such parameters as packaging assume a higher percentage of the value added in product design, the requirement for more comprehensive data handling of the actual graphical descriptions grows. Design systems evolve to support the required design methodologies and become sufficient to deal with parameters such as the complexity of the packaging variables.

The requirements for graphics in some cases may not be as critical since the application can be handled with stick figure representations and the graphics referenced at the last step of a data path.

In products such as electronic circuit boards there is a demand for customization which may involve the definition and manipulation of graphics data. The key factors requiring this capability are variable part placement for various configurations using the same outline and the associated requirements for part distribution. There may also be a requirement for expansion in the number of system variations, and the ability to design a physical package which can be adjusted for various potential conditions in the manufacturing process.

A significant difference may exist in various products depending upon the repair and maintenance procedures. Faults or errors which are detected, either in design or process, are often required to be repaired quickly to provide timely service and minimize costs.

The organizational model is always subject to change in order to handle new product structures of fabrication techniques. These types of occurrences require modification to the data structure to add the new information. The data content can be used to dynamically define the data element and physical record relationships for each file. The application software can be used to interpret the fields to drive some of the processing routines. In this manner, the length and relative displacement of data can evolve over time with minimum impact on the existing software base.

Correlation to the administrative structure is accomplished by allocating the data to specific levels of a bill of material hierarchy. Part numbers are assigned to the hierarchy levels based on the necessary material logistics and data identification requirements. References to engineering specifications ensure that the product is built to an approved process. The part number entries are per the coding of the release rules using a unique formal identification part number. The same data structure also resides in the administrative system. Prior to release to manufacturing, the structures are compared. Any discrepancy is corrected prior to release. Manufacturing defines the bills of material from the design structure. The timing of release is accomplished by communicating with the product development organization, based on their development and engineering change schedules.

PHYSICAL DATA STRUCTURES

A complete description of the data elements and physical records should be maintained by the design automation development group. This library contains all the elements that are part of the design structure. A subset of these are mapped to the manufacturing technical data software interface as summarized below:

Product description—Outline
Bills of materials—Part numbers
Graphics specification—Shapes
Documentation—Assembly/outline drawings
Test specifications—Functional/diagnostic

The following describes the basic function of each record and indicates the implied cross-references to other records and data elements. The intent is to provide the conceptual relationships required for the data structure.

Control Records
    Field format   Data elements, type, and length for this interface.
    Record format   Describes the records written on the interface and the fields.

Administrative Records
    Assembly description   Identifies the units' owner, location, release dates, design coordinates, design quadrants, and co-ordinate increments. This record can function as the file header for data processing.
    Released-part numbers   Defines and tags the levels associated with the assembly and subassembly part numbers being released.

Used-part numbers	Defines the previously released assembly and subassembly parts used by this design.
Design Details	
Shapes	Records the type and characteristics of the shapes, utilizing graphics language.
Shape models	Provides the ability to define data for multiple-use design entities.
Shape model reference	Defines the data transformation to place a shape model on the assembly.
Unused nodes/shapes	Defines geometry that is not uniquely associated with a part.
Connections	Defines the interconnections which are implemented in the physical and electrical design.
Test	Enables programs preparing data for test operations.
Test Data	Raw data which are converted to test hardware stimulus response data for testing the product and diagnosing errors.
Documentation	Image information which manufacturing will route to the appropriate plotter for output; it may be in a format derived by design automation applications and is not usually manipulated by application programs.
Notes	Provides special instructions with minimum syntax for readable instructions to manufacturing.

APPLICATION DEVELOPMENT CONSIDERATIONS

Design automation and process automation application development will be dependent on the following areas which will affect the interface:

1. Data capture systems and procedures utilizing interactive workstations to improve the data preparation process.
2. Changes in the release action which prompt changes or corrections in the manufacturing and administrative structures.
3. New equipment adaptable to the needs of more complex products, and software links to commercial equipment.
4. Expanded graphics which allows for more flexible design rules.
5. Programs aimed at reducing the time it takes for manufacturing engineers to analyze a design for manufacturability and relay their recommendations to the designers.
6. Testing strategies which enable a defect to be quickly diagnosed by the relationships defined on the interface.

7. Overall efforts to improve data quality and minimize error conditions, which cause retries and lost development.

APPLICATION SOFTWARE CHARACTERISTICS

Application software acts to hold the centralized factory together. There are basically three kinds of automatic inspection and testing software: (1) the self-test, (2) the self-learn, (3) the simulator.

Self-test software can be used at several levels, with various portions of the computer hardware dedicated to the testing. An advantage of self-test software is that it can be written by the same designer responsible for the unit to be tested. The result is usually better than could be achieved in most cases by the generic type of test equipment unless it is particularly well suited architecturally to the task and programmed carefully.

Self-learn software is most useful for simple go–no-go tests such as the detection of electrical shorts and opens. The algorithm for shorts measurement is straightforward and allows this technique to be practical even for complex equipment.

Simulation software for electronic equipment can be grouped into two categories. Automatic program generation (APG) decides on the input stimuli and estimates the fault coverage for the program. Simulators take a given input stimulus and calculate the output response.

STRUCTURED SYSTEMS

The single biggest problem in the successful implementation of computerization throughout the factory may be the long system development cycle. As hardware becomes more powerful and more versatile, the demand for supporting software grows and there is an inability to keep up with these demands.

The structured programming techniques of the '70s can improve the productivity of system development by providing guidelines which facilitate the processes of design, code generation, and maintenance. Languages such as Pascal and C are far more suited to structuring than, for example, FORTRAN.

STRUCTURED DESIGN

This method is based on measures, analysis techniques, and guidelines for following the flow of data through the system to formulate the design. It might be used in the following form for implementing a factory automation system. The data flow is found for each data transformation while noting the transforming process and the order of occurrence. A system specification is used to produce a data flow diagram, the diagram is used to develop the structure chart, the structure chart is used to develop the data structure, and the results are used to reinterpret the system specification. The process is iterative but the order of iteration is not rigid.

In structured design the key task is the identification of the data flow through the system and the transformations that the input data undergo in the process of becoming ouput. The process appears simple enough, but consistently identifying the transformations of data can be difficult. It is possible to have too little detail in some parts of the data flow and too much in other parts.

Identifying the incoming and outgoing flow boundaries is important in the definition of the modules and their relationships. The boundaries of the modules can be moved, leading to different system structures.

The structured design method helps in the rapid definition and refinement of the data flows. This has been done for some complex control applications. But the techniques used are not always an integral part of the structured design method. The method and its graphics can also reveal previously unknown properties of some systems such as the generation of information already contained elsewhere in the system.

The method is well suited to problems where a well-defined data flow can be derived from the specifications. Some of the characteristics that make the data flow well defined are that input and output are clearly distinguished from each other and that transformations of data occur in incremental steps.

A software design method is a collection of techniques based upon a concept. Some other design methods include (1) the Jackson method, (2) the Logical Construction of Programs, (3) the META Stepwise Refinement (MSR), and (4) Higher-Order Software (HOS).

Each of these methods prescribes a set of activities and techniques intended to ensure a successful software design. The basis of either the Jackson Method or the Logical Construction of Programs, which is also called the Warnier Method, is that the identification of the data structure is vital, and the structure of the data can be used to derive the structure and some details of the program.

PROCESS CONTROL

The application of factory automation to process control can be grouped into two areas: continuous processing and discrete-parts manufacturing. Continuous processing is concerned with the operation of valves, motors, and other controls in response to measured variables such as temperatures, pressures, and flow rates (Fig. 10-2). Discrete-parts manufacturing is concerned with the positioning of machine elements for fabrication purposes such as metal cutting or metal forming operations. Computers are used for the positioning of the cutting tools and the controlling of the machine motions as well as for the transfer of parts

250 TRANSDUCERS FOR AUTOMATION

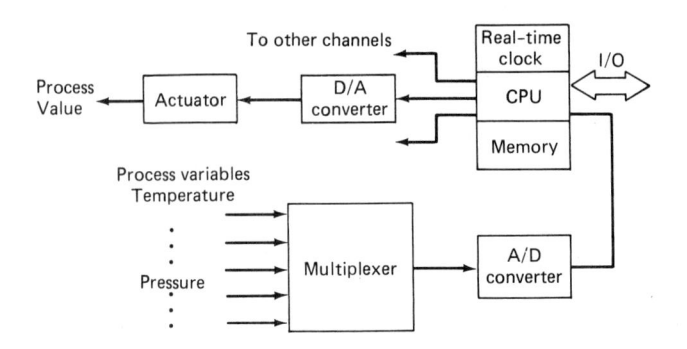

Figure 10-2. Process control.

and materials among work stations. One of the major applications of computer technology in discrete-parts manufacturing has been numerical control, which provides the means used to position tools and machines along prescribed paths. This same technology is also used to control robots in many industrial applications.

These applications use a wide range of computing power. There are the large mainframe computers used in large refineries and chemical plants to control the petrochemical processes. Large mainframe computers are also used as control centers for many direct numerical control (DNC) systems. At the other end of the industrial control system are the programmable controllers (PCs). Early PCs were limited to replacing relay panels or timers, but now most PCs are microprocessor-based and are used for elementary process control. There are also a wide variety of mini or microcomputers. The 32-bit superminicomputers that are now available overlap the domain of the large mainframe computers in many aspects.

A simple process control system might maintain the temperature of a process tank which is heated by steam. The analog temperature signal is converted to a digital signal and compared with the commanded temperature, which is also a digital signal. The difference between the two is the error signal, which is converted back to an analog signal, amplified, and applied to the motor. The motor then adjusts the valve to change the flow of steam, forcing the temperature toward the desired value.

Conventional process control systems are also involved with the collection and processing of analog sensor data into digital form for the following purposes: (1) processing to obtain additional information, (2) storage for later use, (3) transmission to other locations, (4) display for analysis or recording.

The data processing requirements might range from simple value comparisons to complex calculations. One might be interested in collecting information, converting data to a more useful form, using the data to control a process, performing calculations, separating signals from noise, or generating information for displays.

The data could be stored in raw or processed form, they might be retained for short or long periods or transmitted over long or short distances, and the display could be on a digital panel meter or a cathode-ray tube (CRT) screen.

Many data acquisition configurations are possible and there are a number of considerations involved in the choice of configuration, components, and other elements of the system.

Both process control and machine tool (Fig. 10-3) drive systems are conceptually similar in that they are based on closing a control loop using a feedback signal derived from the function or device being controlled. A closed-loop control system compares the behavior of the device being controlled with the behavior commanded; if a difference or error between the two exists, it automatically produces a correction signal which tends to drive the error to zero. If a machine tool drive is commanded to move the machine slide to a certain position, until the slide reaches that position, the position feedback will differ from the input command, and the actuator will continue to drive the slide toward the commanded position.

Industrial computer system configurations have evolved from the basic scheme where a single computer performs all of the supervisory control and process monitoring. Here, the computer monitors a process by reading analog or digital data from the input/output equipment. It may change a control set point in performing a supervisory control function.

The control console can consist of CRT displays, input keyboards, and printers. The peripheral equipment is connected to the I/O channel by the interface circuitry. This hardware performs most of the functions required for the operations between the I/O channel and the peripheral device, such as address detection, decoding, timing, and error detection/correction.

The single/central computer configuration has had a major influence on the technical and operator aspects of computer control. The central computer started with several distinct advantages. A larger variety of more powerful software has been available for the large computers than for small systems. The larger machines had an advantage when a complex such as a fast Fourier transform or

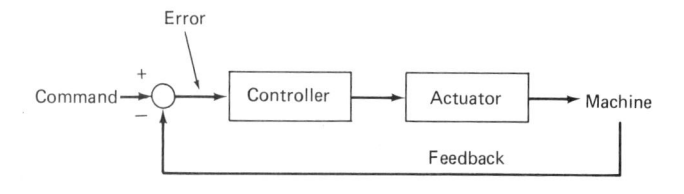

Figure 10-3. Machine control loop.

the solution to a difficult control algorithm was required. Process optimization can require many calculations and a large memory capacity, which made the larger computers more suitable. Control functions which do not require such complex calculations can be handled by smaller systems. Many of the control tasks required in discrete-parts manufacturing involve simple sequencing depending on binary inputs and outputs. These tasks can be performed using a microprocessor or programmable controller.

MULTIPLEX CONFIGURATION

There are several configurations that can be used for the multiplexing system in process control or discrete-parts manufacturing. In remote multiplexing, the multiplexer units are spread throughout the system and the analog and digital signals are sent to the nearest remote multiplexer. A/D converters in the remote multiplexer convert the signals to digital words which are usually 16 bits long.

A control unit signals the remote multiplexer and requests the unit to send a particular block of words or a group of blocks. The remote multiplexer then responds by sending the data requested to the control unit. The control unit will scan the data to check that the transmission is complete. After the data are checked, they may be sent to the central computer or displayed. Remote multiplexing can also be used in systems where there is no central computer. The use of remote multiplexing has increased since it can reduce the cost of the wiring required in a process system.

Remote multiplexing can also be used in systems for multiplexing analog signals. This requires a less complex multiplexer.

Remote multiplexing is reliable enough that it can also be used for control signals. This type of system is sometimes called a total multiplexing system. Both analog and digital signals flow either to or from the computer and the remote multiplexers.

Most remote multiplexing systems transmit data at a rate which allows the update of each analog value at least once every second. This is fast enough for most petrochemical applications, since process values which are updated each second appear to be quasicontinuous.

PROCESS SYSTEM CONSIDERATIONS

To accommodate the input or sensor voltage in the process control system some form of scaling and offsetting may be required to be performed by an amplifier. In order to convert the analog information and if it is from more than one source, additional converters or a multiplexer will be required. A sample-hold can be used to increase the speed at which the information may be accurately converted, and a logarithmic amplifier may be required to compress analog signal information. If the transducer signals must be scaled from millivolt levels to an A/D converter's typical bipolar 10-V full-scale input, an operational amplifier will be required.

When the system involves a number of sources, each transducer can be provided with a local amplifier so that the low-level signals are amplified before being transferred. If the analog data are to be transmitted over any distance, the differences in ground potentials between the signal source and the final location can add additional errors to the system.

Low-level signals may be obscured by noise, rfi, ground loops, power-line pickup, and transients coupled into signal lines from machinery. Separating the signals from these effects can become important for system operation.

Most systems can be grouped into two characteristic types of environments. These are the more benign laboratory or office type of environment and the more hostile factory or outdoor mobile environment.

The latter includes industrial process control systems where temperature information may be developed by sensors on tanks, boilers, vats, or pipelines that are sometimes spread over miles of facilities. The data are often sent to a central processor which provides real-time process control. Steel and petrochemical production and machine tool manufacturing are characterized by this environment, in which vulnerability of the data signals may require isolation techniques or intermediate buffering. The hostile environment may also demand wide temperature components, shielding and common-mode noise reduction techniques, data conversion at several intermediate points, and even redundant circuits for critical measurements.

In the laboratory environment, data collection may take place under more favorable conditions, but the measurements may be much more sensitive and protective methods may also be required.

POSITIONING CONTROL

The trends in the development of distributed systems and the use of microcomputers as control elements have resulted in a wide variety of positioning control system designs. Early controllers have evolved from programs stored in time-shared medium-sized computers. This was followed by the use of minicomputers and then special-purpose hardwired logic to the current compact designs which employ microcomputers. The common objective of these controllers has been to provide satisfactory dynamic control for inputs which are in the form of discrete signals. The basic dynamic characteristics of the power drives were usually modeled after a Type 1 servo: a double integration with velocity feedback and velocity and acceleration limiting. This covers most electrical power drives and many electrohydraulic power drives.

The usual design approach is to make the model of the controlled variable appear to have the same form from the controller's point of view. This is usually done through the selection of the velocity loop gains. The result, when done properly, is a controller that can be used with a wide variety of power drives.

In many distributed systems a major consideration is to avoid commands from the master computer at an excessively high rate since high update rates can saturate the bus bandwidth. This usually limits the sample rate to about 20 to 30 samples per second. At a low rate such as this it can be difficult to maintain smooth system operation.

A conventional analog system driven at 20 samples per second through a digital-to-synchro converter can lead to rough, uneven operation of the power drives, resulting in premature equipment failures.

The usual solution involves some form of data extrapolation for digital conversion. The discrete low sample rate can be modified and sent to the analog loop in the form of a quasicontinuous signal. Then the control loop can be closed digitally and a digital error signal used. A sample rate of 20 samples per second can then result in acceptable performance with position feedback data from a shaft encoder or similar device.

While the earlier controllers used a variety of circuit elements, microprocessors are used almost exclusively today. These controllers generally employ four types of functional sections or modules.

The processor module provides arithmetic capability, temporary storage for calculations, and the mode select and timing and control functions. All of the calculations are usually performed by this module.

The interface module provides the signal interfacing requirements for the different types of systems to be controlled. The characteristics of these systems, such as their acceleration limits, may differ; unique coefficients for difference equations and the solution of various polynomials may be required. The values of these coefficients are usually stored in ROMs.

The analog output module contains the buffer amplifiers and the D/A converters required for the conversion of the digital commands from the controller into the analog signals needed for control. Each output of the D/A converters usually has a connection available for a separate summing amplifier so that it can be summed with the output of the feedback device.

The status test module is available to generate input commands for test purposes. A number of commands, such as steps, ramps, and sinusoids may be generated. The coefficients for these functions are stored in ROMs.

Stepping and servomotors are used in applications which require precise motion to reach a digitally defined position. The commands are supplied to a stepper or servomotor in a closed-loop system and the typical approach is to use an acceleration to the operating speed and then a deceleration to the final programmed position.

Digital interpolation techniques can be used to generate command pulses for a linear velocity ramp to minimize the travel time to the desired position. This allows the last motion pulse and zero velocity to occur simultaneously and reduces the lag between the command and actual positions to provide faster positioning.

Another technique is to use an exponential velocity change for acceleration and deceleration. A closed-loop system with velocity and position feedback produces the exponential output velocity change from a velocity step input command. For a stepping motor control, a voltage-controlled oscillator which is coupled to a pulse generator will produce either an exponential rising or falling voltage for the acceleration or deceleration. In a servomotor system, feedback with an exponential characteristic can be used to produce a lag between the commanded and actual positions. This lag will provide a deceleration and prevent overshooting the commanded position.

In an open-loop system aging and temperature effects in the analog circuits may cause variations in the speed and the exponential time constant. If the deceleration is started at a fixed distance from the final position, the part being positioned may stop short of the destination or arrive at too high a speed and overshoot the position. A lower final speed can be used for the final section of the approach but this may be too slow.

Another technique for stepping motor control is to input a linear voltage ramp to the oscillator producing the command pulses. The positioning time is faster compared to the exponential method, but a switch to a lower final speed is still required. A microprocessor system can monitor the command pulses, determine the acceleration and deceleration phases, and decide when the programmed distance is reached.

PROGRAMMABLE CONTROLLERS

Recent advances in microprocessor technology now allow programmable controllers (PCs) to be used for increased applications in both processes and discrete-parts manufacturing. The quick growth and acceptance of early memory-based programmable logic controllers was due to their simple ladder-diagram programming. This is still a very popular programming technique for PCs, but newer products offer English-language programming that is done with functional blocks.

The newer PCs can handle floating-point calculations or compute trigonometric functions as well as square roots and other transcendental functions. They also have the capability to manipulate relatively large blocks of data.

They are becoming an important part of factory automation, serving as links between machine tools, processes, and higher-order systems. The larger PCs can perform functions in process control that formerly required dedicated computers. These PCs can handle simple process equations and correct process variables in real time.

PCs are used for batch tracking to ensure that the components of any one batch are uniform. Report generation is another area being exploited in factory automation. Many plants are also using PCs for motion control and even though they cannot completely replace NC equipment, PCs can perform many simple tasks in motion control such as positioning parts for assembly.

Motor control is another important function of PCs. They can be programmed to deliver specific numbers of pulses to drive stepper motors, and with the addition of I/O boards they can drive servomotors. The PC generates the initial command, and a separate processor on the I/O board handles the control function and performs the continuous position corrections. Some microprocessor-based PCs provide closed-loop point-to-point servopositioning. An interrupt scheme is used to divide the operation of the microprocessing between the ladder logic scan and the motion-control servo. This allows the PC to service the servoloop frequently enough to provide the precise positioning required.

The higher-level PC's are not powerful enough for contouring applications, but they can be used for point-to-point applications on production equipment such as grinding machines.

ROBOTIC SYSTEMS

A new level of machine intelligence and independent decision-making capability is planned for use in future automated factories. NC and other factory systems are expected to take on functions far beyond the simple execution of machine-control instructions. These systems will sense and adapt to changing cutting conditions without human intervention. Adaptive control will automatically detect and replace damaged cutting tools and compensate for temperature variations and excessive workpiece vibrations.

Automated manipulator arms or industrial robots will be used to perform many material-handling functions in factory systems. Robots can select and position tools and workpieces for NC machine tools.

The hands of robots, which are called end effectors, are used to position and operate tools such as drills and welders. The instructions for controlling the joints of a robot are automatically calculated by a computer to produce the required motion. The workpiece orientation is changed through the action of the robot's joints.

Most robots have multiple joints (See Fig. 10-4) which are manipulated to produce the required robotic motions. Usually each joint of the arm is considered as an independent system for position control. Although a computer forms part of the system, it may only provide the set points for the servo controllers. The control loop often uses a dual feedback path, with an internal loop for the arm dynamics.

Advanced robot systems use automatic programming to determine the joint commands. These robots will usually repeat sequences of programmed move-

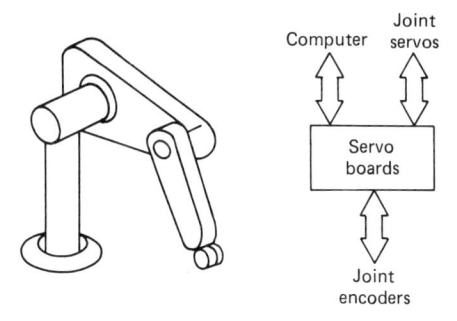

Figure 10-4. Robot system.

ments within a specified repeatability and a reasonable precision. They are suited to repetitive tasks that do not require high accuracy such as welding and spray painting in the production of automobiles, appliances, and other high-volume products. This type of application in mass-production operations usually allows the robot to pay for itself in less than two years.

The more advanced robots use innovative vision and control systems. The control techniques may use feedforward cancellation for gravity effects and adaptive control using force sensors in addition to motion sensors. A system of microcomputers with high-speed sensors for interactions with the environment may control the robot. A parallel processing structure in the control system and real-time decision-processing software for the sensor system allow the sensor data to be taken at high speed. The robot may then sense changes in its environment such as sudden shifts in the position of a part and react to them in the proper way in order to accomplish the required task.

In a typical automatic, vertical-spindle, numerical control application, a set of sensors would measure factors such as speed, temperature, and tool position. These feed the microcomputer system with simultaneous channels of information which is analyzed using fast Fourier transforms for a dynamic analysis of the machine's performance. A memory map may be used for the machine's systematic errors, such as errors in the positioning of the tool or table. These are then corrected as the machine operates.

The control system must interact in near real-time besides being user-friendly. A simple, English-like command language will make it easier to program the robot quickly. The early robots used manual teaching, where the user leads the robot through the required motions. These are then recorded in the robot's memory. These were usually part transfer and other point-to-point applications where only the end points of motions are recorded since the motion path and velocity are not critical. For applications such as arc welding and spray painting, the entire path of the robot arm may be recorded.

Manual teach programming requires only a small amount of training which can usually be done by shop personnel. The cost of the robots and the normal maintenance required is usually less than that of the more sophisticated robots.

Teach programming is too time-consuming and error-prone to make it suitable for most complex processes. Production facilities are usually down during the programming process. Modifications to accommodate design changes or new tooling can require the entire program to be redone.

Off-line programming does away with most of these disadvantages. In this technique the user describes the required sequences of movements and operations through a computer terminal instead of the robot hardware. Robotic languages have been developed for this purpose.

Off-line programming defines and documents the robot instructions better than manual teach programming. The available software aids make the programming task faster and more accurate. Subroutines can be developed for repeated steps and then connected together to build the program. Modifications are done at the keyboard using an editor and symbolic data references to connect the subroutines. Production is not held up and off-line programming lends itself to more flexible use of the sensor data as well as to adaptive control.

THE USE OF LASER MACHINE TOOLS

Lasers have been used since about 1965 with materials-handling and numerical control systems to provide automated processing in welding, cutting, drilling, and heat-treating applications. Laser machine tools allow a number of advantages. These noncontact tools operate continuously, eliminate wear and breakage, and are more energy efficient. Lasers can provide a solution to problems in the fabrication of products made of metal, plastics, rubber, ceramics, glass, paper, and wood. Laser metal-cutting systems can be combined with punch presses, moving coordinate tables, and numerical control systems. The laser is an excellent tool for cutting lines at an angle and other contours including complex outside shapes. Narrow slots, angled slots, and cutouts with acute angles can be cut by lasers. Many high-tensile-strength materials that are difficult to punch can be cut by lasers. The edge finish produced in the laser cutting of sheet metal is superior to that obtained by contour nibbling or plasma cutting. The edge is relatively smooth and square, the kerf width is narrow, and the dross is minimal. The heat-affected zone next to the cut is very small, which facilitates the finishing, forming, and joining operations. CO_2 laser welding can be performed on such materials as cold-rolled steel, titanium, stainless steel, and alloy steels. Aluminum, copper, and similar metals along with their alloys are usually unsuitable. Laser welding often can be the only practical method for bonding dissimilar metals. Laser welding is superior to plasma arc welding since it is stable and controllable. Its energy is focused into one-tenth the diameter with minimal heat damage to adjacent material or devices. An inert gas shield must be provided to prevent oxidation of the material being welded.

Ferrous alloys can be laser-welded to 0.08 in. (2 mm) penetration. The advantages include:

1. Minimal heat input to the part.
2. Welding in a jet of inert gas at atmospheric pressure.
3. No filler material.
4. Minimum fixturing since it is a noncontact process.
5. Minimal distortion area.

Electronics manufacturers use lasers to drill and cut ceramic and other materials. The laser is used here as a scribing tool and connected to numerical control equipment. The 10-μm wavelength of the CO_2 laser is absorptive and has enough peak power for complete vaporization. It also allows high repetition rates for a high throughput.

Laser scribing can produce a cost savings in many manufacturing processes. Multibeam output capability is an important factor. Instead of delivering the entire output in one beam, the beam is divided into two, three, or four beams, with each directed on the workpiece to increase the throughput. A scribing laser can be operated in the pulsed mode with an intermittent output of beam power using a scribe width of 0.006 to 0.008 in.

Lasers are also an aid to manufacturing engineers in increasing productivity. As a highly efficient energy source, they are easily adapted to automation for micromachining control.

NUMERICAL CONTROL

One of the most mature of the automation technologies is numerical control (NC). This is the technique of controlling machine tools with prerecorded, coded information.

These automated NC machines drill, grid, cut, punch, mill, and turn raw material into finished parts. The technique was developed in the '50s to make contoured parts for aircraft. Today it is used in many other industries and these automatic machine tool systems have become much more refined.

The basic NC system uses programmed instructions which are stored on punched paper tapes. The machining instructions are interpreted by paper tape readers connected to the machine tool. Advanced systems use computer numerical control (CNC), in which the machine is controlled by a dedicated computer with the NC instructions stored in its memory. This allows the instructions to be stored, handled, and changed more efficiently. Diagnostic and other intelligent functions beyond simple machine control may also be performed by the computer.

More complex automated systems use direct numerical control (DNC), which is also known as distributed numerical control if the machine control is distributed among a network of computers. In these systems the individual manufacturing units are connected through communication lines to a central control computer. These communications lines can also be used for the feedback of production and machine-tool status from the shop floor. Some of these systems use a hierarchical arrangement of computers connected to a central computer while others have direct interfaces between the central computer and the machine tools.

Two basic approaches to distributed control are the loop approach and the unit approach. The loop approach relies on satellite processors to perform specific functions. A single processor might perform a specific machining operation. A single processor might also perform a single function in several different loops. But if a processor is dedicated to only one loop, then a single failure will cause the loss of only that loop. The unit approach relies on a separate control system for each unit or cell in the factory with a control processor assigned to each unit.

Some advantages of distributed control can be obtained by either approach but others can only be realized using the unit approach. A major advantage of distributed control is the increase in efficiency. Since the processor's I/O equipment is located nearby, most wires must only be run a short distance. This provides a savings and since the runs are shorter, there are fewer problems from interference with the low-level signals. There is no need for a large control panel with distributed control. The unit is inherently modular. Computer control can be added to the units in a plant one at a time, without disrupting control of those units already on computer control. This is attractive in a plant operating without computer control. The most likely machines for computer control can be established after a study. The unit or cell may be a production bottleneck, a large energy user, or one which is difficult to control.

The instructions for these NC systems can be prepared either manually or with computer assistance. In manual programming the tool-path coordinates are first calculated from the part dimensions. Then the program is prepared in an alphanumeric format which is required for entry into the machine control unit.

Manual programming provides a large degree of user control over the NC instructions, but the numerous calculations are tedious and prone to errors. The programmer must have a knowledge of the particular control language so the instructions are entered in the proper format. Extensive measurement of the part is needed in order to define the program sufficiently to produce a properly machined part. Lengthy trial machining may also be required to refine the program. Manual part programming is typically used in small machine shops.

In computer-assisted systems, the programming language acts to extend the user's capabilities and the NC instructions can be developed more quickly and with fewer errors. The instructions typically are written in a language like APT or Compact. The instructions relieve the programmer from manually entering any geometric data.

APT is the older and more universal language. It has become the standard of the NC industry. The languages allows the user to produce NC instructions from English-like statements. Working from an engineering drawing, the part programmer enters statements identifying the machine parameters, the part shape, and the tool type. The computer then makes the calculations defining the tool path to produce the shape for complex curved surfaces. For complex curved surfaces the user defines the control points and surface equations. When numerical control (NC) instructions are written manually, the process tends to be error-prone because a large number of calculations are required to define the tool paths.

In a typical computer-assisted system, when the NC program is selected, a menu appears on the screen, and the user designates the machine that is to cut the part. When the machine is selected, other menus are used for the selection of

CUTTER SIZE
TYPE OF MATERIAL
TOOL OR PART SPEED
MATERIAL THICKNESS
CLEARANCE THE TOOL MUST HAVE WHEN NOT CUTTING

The screeen will then display a geometric model of the part on which the programmer defines the machining operations using different types of lines. A dashed line might indicate a rough cut while a solid line might indicate a finish cut. The system uses these inputs to automatically generate an NC tape to machine the part. The system allows the programmer to create the NC instructions graphically, without requiring a detailed knowledge of programming languages.

Most of the NC programming packages can also simulate the tool paths so that the program can be checked. The tool motion may be animated to allow the programmer to observe the tool as it moves on the part.

Consider the tool motion required to machine a three-axis composite surface of slanted planes and cylinders. This can be simulated on the computer system. Then, after the operator verifies the tool path, the system will automatically generate the APT language instructions for producing the part on the NC machine.

The motion of a five-axis machine tool can also be simulated. The front, top, side, and isometric views of the path can be displayed for program verification and editing before the NC instructions are produced. The motion shown on the screen can be slowed and one can zoom in on a specific area to clarify details and increase accuracy. These features allow a verification of the proposed program.

After the program is checked, the computer translates or postprocesses the tool-path description into a set of coded instructions for the specific machine. These instructions may be stored on punched paper tape or magnetic tape for input to the machine tool or they can be transferred directly to a memory in the factory control system. Computer-assisted systems can perform many activities automatically. Some systems use the geometric data created by computer-aided design techniques as a basis for producing NC instructions which in turn calculate such machining parameters as (1) optimum feed rates and spindle speeds, (2) proper sequence of work elements, (3) optimum tool paths to fabricate the part.

After the processor develops the NC instructions, the process can be simulated as an animated color solid model on the screen of an interactive graphics terminal. Using this display, the programmer can check the validity of the machining process and then use the terminal to make any necessary changes.

In addition to computer-assisted programming, the software may use the geometric model created in computer-aided design as a basis for producing the NC instructions. The geometric information is accessed through a shared data base in the computer; thus, the programmer does not have to enter data manually. The computer may also prompt the programmer to respond to questions displayed on a terminal screen.

In a fully integrated factory system with a common data base the processor can extract virtually all of the information required for generating NC instructions. The part descriptions can be obtained from the geometric modeling portion of the data base, and the appropriate machining parameters can be taken from process planning data.

Using this technique the processor can (1) automatically recognize the solid model of the part, (2) identify the material to be removed from the raw stock to develop the part, (3) select the tools required to produce the part.

An advantage of computer-based programming systems is the speed with which they can generate the NC programs. The programmers need not be familiar with the intricacies of tool-path generation.

In some systems, many of the machining parameters such as cutter types and feed rates remain constant so that the programmer input beyond the geometric mode is minimal for simple part shapes.

SOFTWARE SELECTION

The software programs available for NC programming may have a wide variety of strengths and prices. Some of these may have originally been CAD programs, which were used in specific areas of electronic or mechanical design. These origins may lead to shortcomings in NC applications. Some CAD-based programs may not be able to execute all of the required commands or carry them out effectively. There is also the possibility that the program will not fit the specific NC task.

Resolution is a basic program capability to consider. The surface finish a program should be able to provide is determined by many cost-influencing factors such as part volume, amount of hand labor, and NC programming and machining time.

If there is a single application that dominates the operation, then one may be able to calculate how much might be saved if productivity were increased by an assumed factor. This savings can be compared to the cost of the system. Instead of basing the system on all the capabilities you would like to have and selecting a package that promises the greatest number of capabilities at the lowest price, the system can be keyed to the most dominant production needs. Two basic tests that can be used to determine if a program is suitable for a particular NC application are:

1. Generate a tool cutting path for a complex part that is similar to the component that will be manufactured.
2. Conduct a series of tests or benchmarks which measure specific program capabilities.

The complete benchmark results are evaluated and compared to a specific program's costs, abilities, and limitations. The following are some typical benchmark tests which reflect actual manufacturing needs (Fig. 10-5). These tests employ a particular shape or machining task which may indicate the limitations of the program.

Normal cutting—Cut a hemisphere with a radius while keeping the cutter axis tangent to the surface. This program must be able to maneuver the cutting tool in five axes and produce normal vectors. This is one of the simplest benchmarks.

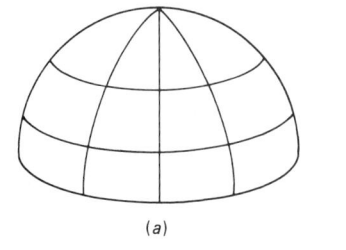

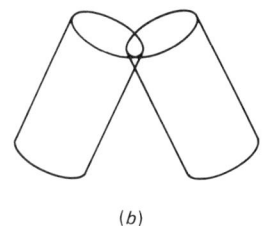

Figure 10-5. Typical NC benchmarks. *a*. Normal cutting. *b*. Intersecting cylinders.

Cut a tapered slot—This test involves cutting a tapered slot with a flat bottom in a block. The taper is the most difficult part since it requires five-axis capability. Some programs are limited to $2\frac{1}{2}D$ machining (x, y, and only vertical z motion).

Cut intersecting cylinders—This test requires cutting two cylinders which intersect at an angle. The difficult part is to maintain a fillet along the lines of intersection without cutting into the work surface. The sequence in which the faces are cut as well as the path the tool takes are important factors in determining the total cutting time.

Cut a raised patch—Patch is faired into the cylinder's surface. The blending of the two diameters is the difficult part since there should be a smooth transition around the patch without cutting into the surface of the cylinder.

Transition cut—Cut a transition from a chamfer to a radius and back to a chamfer. The transition is difficult since it is not a basic geometric shape. The program must be able to generate a sculptured surface to maintain the required smooth transition.

Sweep cut—In this test, one generates a 2D shape and sweeps it through a complex path, creating a channel in three dimensions. The program must have the capability to consistently maintain the 2D cross section.

Tangent cutting—In this test two planes are tangent to a circle and several cylinders intersect the planes and the circle. A section of constant depth is cut with a boundary defined by the centers of the cylinder. The cutting path must be based on a number of tangent points, while maintaining a consistent cut depth in multiple planes and the hole wall perpendicular to its respective planes.

Convex to concave cut—This test requires the program to cut a surface that changes from convex to concave. The geometry of the ends is defined and a smooth transition is needed which requires the program to establish boundary conditions between the defined end points.

SCULPTURED SURFACES

Sculptured surfaces are made up of arbitrary, nonanalytical contours that may not obey mathematical laws and were considered to be impractical to machine with NC. These surfaces are found in a wider range of components including those used for aircraft, automobiles, construction and agricultural equipment, machine tools, and appliances. These surfaces traditionally have been defined by subjective curve-fitting techniques and manufactured with expensive operations.

These sculptured surfaces with their arbitrary nonanalytical contours cannot be defined by mathematical terms. They are typical of aircraft, automobiles, office machines, and computer-terminal enclosures or products which depend on free-flowing surface contours for performance or aesthetic appeal. Part shapes that can be defined by mathematical equations such as planes, cones, spheres, and paraboloids can be readily programmed for NC machining but considerable computer power may be required for the more complex parts. There is no simple way to create an NC part program for machining the contour of a sculptured surface since the surface cannot be defined mathematically. The complex contours can be represented by a network of patches. Each of these may have a mathematical contour that serves as a close approximation to a corresponding surface element of the surface to be machined. These patches are twisted and stretched by the computer to conform as closely as possible to the sculptured surface and blended together smoothly.

There are programs for generalized shapes that can be easily implemented on a wide range of computer systems. Some programs can define arbitrary shapes and manufacture them on NC machine tools. Various versions are being used in industry. Many manufacturers have tailored the program to suit their own particular requirements and use it as a production tool. Some vendors also market modified versions as part of their overall commercial software packages. Here the complex surfaces are represented by a network of interconnected patches in the sculptured-surfaces program. After the surface is modeled, the program output can generate NC machining instructions for a contoured part.

DISTRIBUTED PROCESSING

Most complex plant control systems can be divided into smaller and simpler subsystems. The responsibility of the complete control system is divided up, with one computer controlling the overall flow of tasks. This concept is known as distributed processing. The processors may not be dedicated to any single function. They are assigned tasks by a master processor which operates the total network.

A primary advantage of distributed processing is the division of labor, since this tends to off-load the processor and improve the performance of the overall system. The improved time response generally results in the improved execution of functions since it is no longer necessary to wait for a central computer to become available.

Distributed processing can also result in improved system reliability and failure tolerance. The remote units can allow operation independent of the central computer for short periods of time; thus a short outage in the central computer or a communications link can be tolerated. It is also possible for the central computer to take over control for a limited loss of remote units.

The network can be designed for operation in harsh, noisy, industrial environments and reliability can be enhanced by (1) fail-safe schemes that prevent any single failure in the communication system from bringing the entire system down, (2) redundancy techniques including fully redundant interfaces and cabling, and (3) self-repairing techniques that automatically disconnect failed devices or add in new devices.

Distributed processing also allows a modularity which is unavailable in many centralized systems. When sections of a process are automated, more remote units can be added. In the initial stages there is no need to install a system large enough to meet all anticipated expansion plans.

In a duplexed configuration a primary or master processor performs the control task. If a failure occurs, the backup processor takes over control. The switchover from the primary to the backup computer must be designed so that it results in only a small control deviation. The primary processor could be taken off line, repaired, and returned to service without interrupting the process. The duplexed processors provide twice the computing capacity, since either processor can control the process. This excess power can often be used for other optional tasks. To guarantee that the higher reliability will be achieved, the system must be analyzed for the effects of each failure classification. The software must also be designed to achieve the higher potential reliability of the duplicated hardware.

In a duplex configuration the software must consider when the backup processor should be activated to take over. The primary processor might set a timer in the backup unit on a periodic basis. Failure to set the timer causes the backup unit to assume control and disable the primary unit. This type of timer is known as a hardware or watchdog timer.

Duplexed operation allows the two processors to have special characteristics. One processor could be used to perform fast floating-point calculations and this would be its function during normal operation of the system. In order for the backup processor to keep up to date with the status of the program, it must have access to the current process status information.

If a shared-file system is used, then this information can be made available in the shared-file system by using an access mechanism known as token passing. Token passing requires that each station on the network have the token before the central computer will talk to the node. Once a device has gained temporary control of the system, it communicates with another device by establishing a virtual circuit.

The method used to determine which node has the token may be either deterministic or sequential, where the token is passed from node to node in a strict order. This requirement can reduce the system response too severely for some dynamic process applications.

Primary devices can receive the token and initiate communications. Secondary devices may only respond when queried. Demand devices are secondary devices which are allowed to behave like primary devices only when under certain specified conditions.

The disk files are serial devices which two processors cannot use at the same time. So the contention problem must be solved by circuits or another processor (an arbiter) which may be part of the file interface.

When the system has several processors, the tokens or requests from one processor to the others are posted in the shared file. Then a scheduling program periodically checks to determine if a request is waiting. The request can then be scheduled according to the system priority.

Some of the sensors in the system could be connected to several process I/O subsystems and others connected to only one. The shared sensors might be those which are critical to the process. If one of the processors becomes inoperable, the other processor still has access to the data in order to maintain control.

Another basic technique is connecting the processors through the I/O channels. This may be either a serial or parallel connection. The transfer of information will be based on the data rate of the slower processor.

In these systems, several operating modes are possible. The processors might cooperate in solving a single problem which requires more computing speed than a single processor allows. Each processor will then operate on a portion of the problem.

DISTRIBUTED CONTROL

The increased use of computer networks in data processing applications along with the increased popularity of remote multiplexing has resulted in the wider use of distributed control in the process control industry. Distributed control differs from distributed processing, in that it also makes use of remote multiplexing. The processors communicate with field-located multiplexers and each processor is usually dedicated to performing the same task in an on-line environment. In many process industries, distributed control has been a requirement.

Microprocessor-based computers now allow distributed control to be used for discrete-parts manufacturing where several separate control devices are installed along various points in the production process.

The basic techniques for distributed control include the loop approach and the unit approach as discussed earlier. The loop approach uses its satellite microprocessors to perform a fixed number of functions. A single programmable controller could perform one function for a number of different loops. When a processor is dedicated to a single loop the controllers can be connected by data buses and may operate independently, communicating with each other, or assume temporary control of each other. A processor failure can cause the loss of only one loop.

The unit approach requires a separate control system for each unit in the system. A microprocessor may be assigned to each unit, such as the control of a distillation column. Both approaches provide some of the advantages of distributed control but the unit approach generally has more benefits. A reduction in wiring is one major advantage. Since the microprocessor's I/O equipment will be located throughout the system, the wiring tends to be shorter. The shorter runs cost less to install, and since the runs are shorter, there are fewer problems due to interference. Other benefits include more programming units. Ladder logic may be entered at individual programmable controllers (PCs) which may be augmented by CRT displays for improved visibility. The CRT displays can be used to show several rungs of ladder logic at once.

The reliability and maintainability of a process control system with distributed control is improved. The cost of microprocessor hardware is low, so processors in a unit configuration can be backed up with a spare (duplexed processing); thus, failure of a single processor will have no effect on operation. Enough CRT control consoles must be used in the system that a sufficient number of operators can control the system during critical periods.

Since most microprocessors are normally on a single printed circuit board, the system is easily repaired. It is usually a matter of replacing the suspect board.

The unit approach is highly modular and computer control can be added to the units in the system one at a time, without disrupting control of those units already on computer control. The most likely unit for computer control can be established after a system study is conducted. The unit may be a system or process bottleneck, a large energy user, or one which is difficult to operate without computer control.

Distributed systems will use at least two communications protocols. One is used for the local units and another for the higher levels. A common local protocol is RS232C, but RS422 offers a greater immunity to rfi and the lower cost of a twisted pair of wires. The higher level, which is sometimes called the supervisory net, will probably use a version of the IEEE 802 standards.

COMPUTER NETWORKS

Computer networks are an extension of the distributed processing concept. The differences between a distributed processing system and a computer network often overlap.

Computer networks consist of two or more computer systems which are separated. The separation distance can be a few feet within the same room or several thousand miles between installations connected by common-carrier communication facilities.

Factory automation networks which can link a large variety of programmable controllers, computers, and other devices in harsh factory environments are altering the concept of modern manufacturing and production control. These networks permit rapid on-line feedback from monitoring devices and allow more effective control of information and processes. They also allow many devices to share peripherals, which reduces the amount of hardware that must be dedicated to single applications and cuts cabling costs.

Typical factory networks may connect the following equipment:

    Programmable controllers (PCs)
    Computers
                                CRT terminals
                                Printers
        A/D converters
                Robots
                I/O devices
                              NC equipment
    Production monitoring equipment

In computer networks, the computers are normally loosely coupled and capable of stand-alone operation. The connection between the processors is known as a link. The link provides the communications channel, through a medium such as a coaxial cable or telephone line. Each device in a network is often referred to as a node. The network can be organized in a number of ways, depending on how the control responsibilities are assigned among the processing units.

The master-slave network employs a host processor (the master) which is connected to satellite processors (the slaves) as shown in Figure 10-6. In a master-slave network, a single processor has control and determines which slave computer will operate on a task. Communication between the slave computers is under control of the master. After it is assigned a task by the master, the slave operates asynchronously with respect to the master until completion of the task, or until it requires service from the master. The master-slave configuration is used mainly to provide system load sharing. A master-slave network with multiple slaves is also called a multipoint or multidrop configuration.

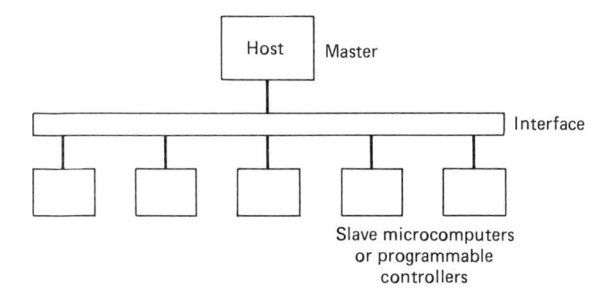

Figure 10-6. Master-slave network.

270 TRANSDUCERS FOR AUTOMATION

The hierarchical network is based on multiple master-slave levels as shown in Figure 10-7. The levels in the hierarchy are assigned responsibilities for specific functions. The highest level in the hierarchy has the responsibility for the major decisions and the lower levels make decisions for the control of specific operations.

Peer or peer-to-peer networks differ in that any device can communicate with any other device and there is no master or bus-arbitration device required. The master-slave and hierarchical networks employ a top-down control philosophy while the peer network uses mutually cooperating computers in which there are no defined masters or slaves.

The peer configuration requires the operating system of each computer to keep track of the status of the other computers in the network. A scheduling program can be used to provide the task distribution. A computer which is busy passes the task on to the next available computer that can execute the task. The time response in peer networks can be difficult to predict, because of the random nature of the scheduling. Peer networks allow access to specialized facilities not available on the originating computer and the processors can use dynamic load sharing to make more efficient use of the total system.

Besides the control configuration, several different physical connection systems may be used for a network. The star configuration (Fig. 10-8) uses a host or master processor in the center of the system. Each professor communicates with the host and any communication between the satellite processors must be done through the host. The protocol is simpler since there is no contention for physical access. The software must control the data flow to and from the host. This is also called a radial or centralized configuration.

A star network operates at the speed of the host, although the transmission speed of a link can be adjusted for the specific requirements of that link. System reliability depends on the host operation, and if all satellites need to communicate through the host at the same time the host may become congested.

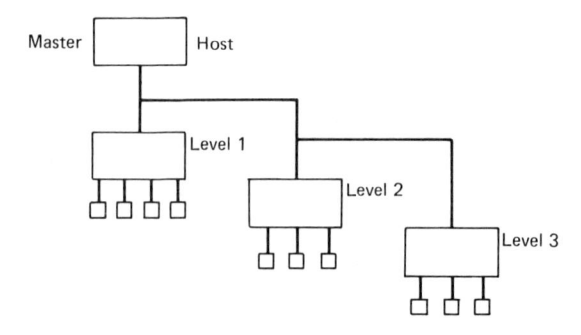

Figure 10-7. Hierarchical network with multiple levels.

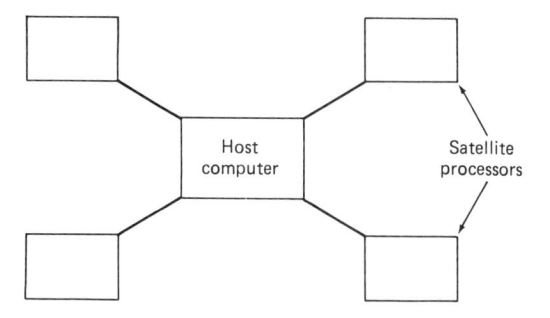

Figure 10-8. Star network.

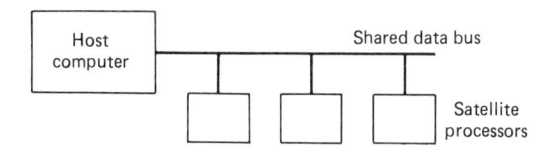

Figure 10-9. Multidrop network.

The multidrop configuration uses a host to control the flow of data between any two nodes. Any satellite can communicate with the host or any other satellite at any one time. This configuration is shown in Figure 10-9 and is also known as a data bus, data highway, or multipoint network. It is easily modified, but there may be unpredictable data rates and contention problems for resources may exist.

In the loop or ring configuration, if a single link breaks the nodes can still communicate. This configuration is normally used in remote multiplexing systems and is shown in Figure 10-10. The loop may begin and end at a loop controller, which is used to control the communications. Messages between computers in the loop are handled as a string of words with some bits or words containing information on the originator and the addresses. When a computer recognizes a message addressed to it, it accepts the message. As a processor

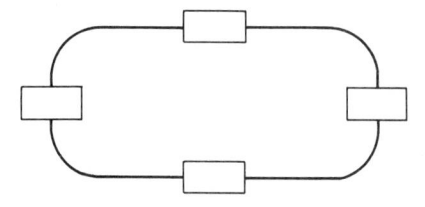

Figure 10-10. Loop or ring network.

272 TRANSDUCERS FOR AUTOMATION

receives and verifies a message, the starting and ending address of that message is passed to the destination processor. This processor then encodes and retransmits the message to the next point in the network. The supervisory processor or controller maintains network information and data link assignments based on equipment conditions, message load, and the most direct route to the final destination. Loops may be difficult to control under some conditions and require higher data rates because of the way the messages pass through the computers. System traffic is generally suspended during reconfiguration.

A point-to-point network requires all processors to have a direct access to every other processor as shown in Figure 10-11. For a network of n processors, $n(n - 1)/2$ interconnections are required. A three-processor system requires only 3 communication links, a five-processor system requires 10, and a ten-processor system requires 45 links.

The advantages of point-to-point include a faster response on lower-grade communications lines and the alternative paths allow messages to be forwarded even if some of the links are broken.

FACTORY NETWORK CONSIDERATIONS

The technology exists today for automating many types of operations in a factory. A comparison of today's automatic manufacturing equipment with the equipment and systems of only a few years ago illustrates the rapid impact of computerization in the factory. But there is much existing equipment, in many plants, that operates as islands which are separate from the mainstream of product design and production. This is limited in view of what is possible in the integration of test instruments and inspection equipment in the manufacturing environment.

As automation spreads throughout the factory, the full potential for increased productivity will not be fully realized unless all of the data inputs are

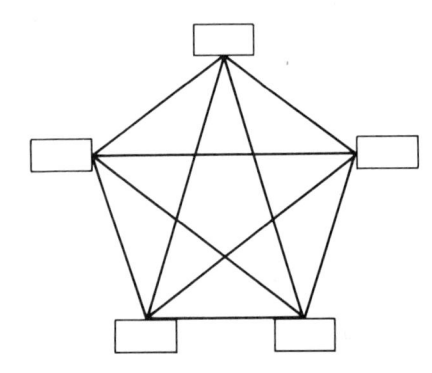

Figure 10-11. Point-to-point network.

automatically acquired and all of the data exchanges occur without human intervention. Inherent in such a system is the computer analysis of manufacturing data to reduce rather than expand paper output. The human operator in these systems should be concerned only with monitoring performance, with an occasional intervention when an unforeseen event occurs.

Any manual data entry is prone to error. It can easily become the plant bottleneck, limiting production and productivity for the entire facility. This can only be eliminated if the input data exchange to all the factory automation equipment is automated.

The test and measurement functions of the system always become more critical as the product reaches the finished product stage. In the case of electronic circuit boards logic analyzers may have been attached to the prototypes and used to run the same test stimulus patterns that were used during the earlier logic simulation phase. The factory system can then capture the outputs available from logic analysis and compare them to the simulation outputs to obtain a comparison. Mixed-mode simulations are also possible with part of the system in hardware and part in software.

In computer-integrated manufacturing (CIM), the automation equipment is likely to be combined with postprocessors, simulators, or other electronic equipment. This allows computer-aided testing (CAT) to take place at a central position in the factory system. The primary interface of the CAT system is with the common data base of design information available from the computer-aided design efforts.

A computer-based factory information system (FIS) can be used to directly monitor the manufacturing process. It ties together the various elements of the factory through computer integration. This provides the basis for total automation. Such a system is complex since it must coordinate many activities in order to maximize productive efforts.

There are several positive benefits to an automatic data exchange in the factory system:

1. It allows the capability to check a product or process design before it is actually placed in production.
2. It allows the critical evaluations necessary to optimize the manufacturing or processing system.
3. Computer control can also allow complete test documentation. This includes the detailed test data, which can be automatically generated.
4. An automated operation similar to that used in process planning can be used to assign the units to the various manufacturing or processing stations.
5. Automating the test interface allows one to react more quickly to changes. But the effective use of computer-aided equipment requires that the system design be essentially complete, since implementing changes results in nonproductive time.

6. The use of appropriate application software allows the manufacturing data base to be automatically converted into codes for such functions as design drafting, numerical control (NC) fabrication, and functional testing. The last operation often requires some form of simulation.

All of these techniques require automatic high-speed data acquisition and data transmission to fully integrate the test and inspection system into computer-controlled manufacturing. The requirement for a factory network includes the computer language as well as the methods of data exchange. There are a number of advantages in the utilization of those methods and procedures that have been standardized or are likely to become standards in the future.

LANGUAGE CONSIDERATIONS

When the first factory automation systems were introduced, they normally used a proprietary language and network. The language may have been an extension of a language standardized by ANSI (the American National Standards Institute); in some cases these extensions were the same as those being covered for inclusion in the next revision of the standard. Often the systems based on nonstandard language extensions did not differ much from those that used a proprietary language, except that they might have been easier to read. Problems occurred because the native language of each computer was different and each needed its own version of the high-level languages to translate program into machine code. In the absence of standards, a program written in BASIC might run only on a machine compatible with the one for which it was designed. The high-level languages have become more and more machine independent.

At first, only logical and mathematical instructions were standardized. Later, input, output, and file handling were added. Program syntax at the command level was standardized while the responsibility for reducing the syntax to native machine language instructions still rested with the computer maker.

In some areas, these standards defined new languages. For automatic test equipment first ATLAS and then ADA was promoted by the Department of Defense. Eventually, the DOD required that all test equipment purchased after a specific date be programmed in ADA. Thus, a test unit designed by a vendor or other party could be used by anyone. This ability becomes more important as the number of different components in the system needing testing increases.

Some of the most common languages used for computer-controlled equipment include ADA, ATLAS, BASIC, and FORTRAN.

If systems use a language of the same form as the one that has been standardized, then the application software can be compatible with a similar system from a different source, if the hardware and software use standard or proprietary interfaces, algorithms, and protocols.

Thus, an important item in the selection of a language for application software is the consideration that must also be given to the languages used in the other networks of the factory system. In order for data to be exchanged between any two networks, we need to standardize not only the language, but also (1) the protocol or the rules for data exchange, (2) the word size or the number of bits in a word, (3) the character set or decimal number assigned to each different alphanumeric character and graphic symbol. The three most popular character sets are ASCII and EBCDIC for alphanumeric characters and IGES for graphic characters.

The networks between which data exchange is desired are connected by an I/O interface. This section may be passive until specifically addressed by a network. It will then buffer the data to be sent to the other network and determine if the data have been correctly received by using a redundancy check.

The interface function becomes more complicated when there are differences in any of the characteristics of the networks being connected. The interface must then harmonize, or eliminate, the difference. One way to do this, for simple differences, is to use programmable read-only memory (PROM). A table look-up can be performed using the incoming bit-stream as the index. The bit-stream required by the other network for the identical meaning is returned from the table. The message is then reconstructed with the header and the trailer of the data stream which were stripped from the incoming message and placed around the new bit-stream.

NETWORK PROTOCOL

In the networking environment, the protocol has come to mean both the format and the set of procedures that allow communications. The layering of protocols is made possible by the stratification of functions among the various parts of the system and allows devices from different sources to communicate as long as they follow the protocol.

As a standard model for a layered protocol for computer networks, the United Nations' Consultive Committee for Telephone and Telegraph (CCITT) has released through the International Standardizing Organization (ISO) its Open Systems Interconnection. This is also known as the Open Systems Architecture and it has been adopted by ANSI.

This important standard defines the function and layering of a set of protocols for the interoperation of equipment built independently. This model consists of seven layers or shells. Starting from the highest or seventh layer, the layers are

> *Level 7, Application Layer*—This layer deals with the specific application such as editing or communications.

Level 6, Presentation Layer—This level provides the techniques for mapping between the various user terminals, which may include special hardware terminals or generic or virtual terminals.

Level 5, Session Layer—This level allows the controlled-access mode to host services such as user LOG IN operations.

Level 4, Transport Level—This level provides such services as flow control, error control, sequencing, and multiplexing.

Level 3, Network Layer—This level contains the set of procedures that allows a host to send or receive data on the network. This includes addressing and network control messages.

Level 2, Data Link Layer—This level provides the line-control procedure that allows the transporting of frames from one computer to another.

Level 1, Physical/Electrical Layer—This level defines the physical or electrical standard such as the type of connector used, the voltage levels, and the function of each connector pin.

Physical communication between the networks occurs only at the physical layer and is managed by the data link layer, which serves the other upper layers. A standard such as IEEE 488 could be used to implement the first layer. The advantage of using standard hardware such as that allowed by the 488 bus is the elimination of the need to design an I/O interface for each instrument. One type of hardware driver, the IEEE 488 controller, provides the handshake, timing, and protocol logic for the instruments. The use of standards such as this has dramatically lowered the cost of inspection and test stations.

The following four major types of transmission media and interconnections are most often considered for computer-controlled inspection and testing: (1) RS-232C/RS-449, (2) IEEE 488, (3) IEEE 802/PROWAY, (4) ETHERNET.

RS-232C and RS-449 are designed for communications between two specific nodes. They do not contain the mechanisms required for transferring data from one selected node to another. These standards have been widely accepted over the years and are found in a great variety of computer and data communications equipment.

IEEE 488 is an instrumentation standard with 24 conductors. It uses a bit parallel, byte serial message. The data input/output lines are either open collector or three state. This allows all devices connected to the bus to listen to messages on the bus and any one device to send messages on the bus. A three-wire handshake system is used to ensure that each device on the system receives all the data sent to it. The handshake system also forces the talking device to operate at the speed of the slowest listener that is enabled on the bus. While this standard has been applied to a broad range of applications, it generally does not have enough of the transmission control features required for many factory control applications. It is useful for the instruments at the basic cell and station levels of the system. For the interface I/O between the test center level and the factory level IEEE 802/PROWAY or another ETHERNET-based standard is preferred.

ETHERNET-based standards use a coaxial cable which is usually less than a few kilometers in length. They can have a treelike or dendritic network configuration (Fig. 10-12), which is a typical characteristic of factory networks. The propagation delays in the cable are short, so transceivers can be used to detect collisions by comparing the outgoing and incoming signals.

There are a number of important differences between these standards, even though they have common origins. ETHERNET systems use a common set of specifications so that every ETHERNET system is compatible with every other one. IEEE 802 allows several options, so there may not always be complete compatability between all nodes of a network, unless each node uses exactly the same options.

Both network standards require transceivers at the nodes, but the cables connecting the transceiver I/O to the node are different. There are also differences in addressing, encoding, control, synchronizing, and access methods.

An access method is required in an ETHERNET-like network since any node many transmit a message frame when the network is available. If two or more nodes are allowed to transmit at the same time, the message becomes garbled beyond recognition owing to the lack of a contention solution.

One way to solve this problem involves employing a Carrier Sense Multiple Access (CSMA) algorithm. One type of CSMA algorithm is collision detection (CSMA/CD). Here, the node is monitored before transmission is allowed. Unless the network is clear, the transmission is prohibited. If the monitoring of two nodes results in a contention, the procedure is repeated using a random timing algorithm. Since the timing is random, the contention is unlikely to be repeated again.

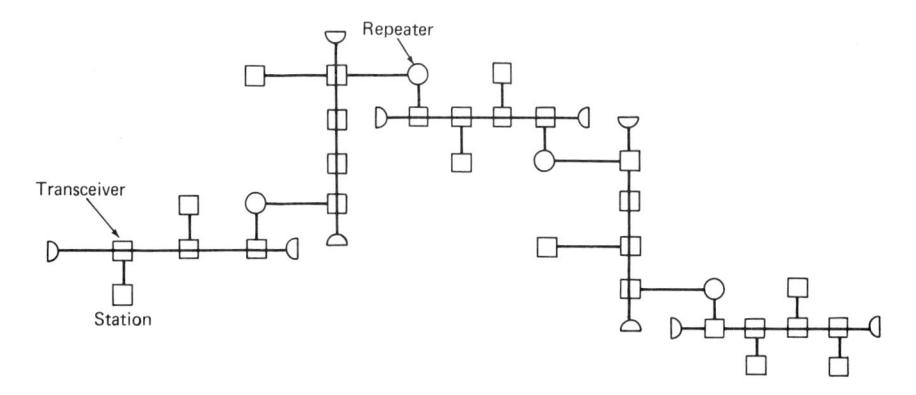

Figure 10-12. Typical ETHERNET configuration.

278 TRANSDUCERS FOR AUTOMATION

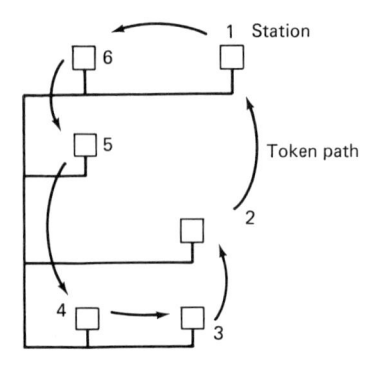

Figure 10-13. Token passing.

The IEEE 802 options allow either CSMA/CD or token passing. The two techniques are not compatible and cannot be used interchangeably.

In token or baton passing (Fig. 10-13), as it is sometimes called, only the node that has the token which is a particular bit pattern is allowed to transmit. The node can only hold the token for a specified time; then it must be passed on to another node. The nodes can be assigned a priority so that one or more nodes has a greater chance to transmit than other nodes.

The CSMA/CD scheme becomes overloaded when the network traffic exceeds about 30% of the channel capacity. Token passing becomes inefficient when the traffic is even less than this.

The major differences between CSMA/CD and token passing are the method of encoding and the need of a carrier for CSMA/CD.

A network could be capable of using either technique and switching between them as determined by a controller. This would require a carrier and under high traffic conditions might not eliminate all contention problems, but it would improve efficiency.

The link layer is concerned with the assembly of data, addresses, and acknowledgments into the message frame. The outgoing frames include the destination and source addresses as a header and an error detection/correction code as a trailer. For received messages, it is in this layer that the source address is inspected and if it is not intended for this node, the frame is discarded.

HDLC PROTOCOL

The protocol at the data link layer is often based on the High Level Data Link Control (HDLC) standard and it supports the higher layers. This standard was developed by ISO.* It was derived from the Synchronous Data Link Control

---

*High Level Data Link Control (HDLC), DIS 3309.2 and DIS 4335, International Standardizing Organization, Geneva.

(SDLC) procedure and is also known as the Advanced Data Communications Control Procedure (ADCCP).

The basic mode of operation in HDLC is an asymmetric primary-secondary mode in which one unit issues commands and the other responds. There is also an asychronous balanced mode (ABM) in which both sides act, each issuing and responding to commands from the other.

The basic frame format for HDLC is shown below:

> FLAG
> ADDRESS
> CONTROL
> .
> DATA
> .
> FRAME CHECKSUM
> FLAG

A frame is flagged to the receiver using an 8-bit flag sequence. Hardware is used to protect against the unintentional signaling of a flag sequence. A storage register is used to hold the preceding 16-bit checksum, which is compared against the checksum that the receiver generates as the frame arrives. The flag protection and checksum verification procedures are available in the form of integrated circuit chips.

Other features of HDLC include error control, sequencing, and flow control. There is also a link initiation and disconnect technique, which uses a set asynchronous response mode (SARM) command, an unnumbered acknowledgment (UA) response, and a disconnect command. A poll/final (P/F) bit is used in command frames to request a response and in response frames to acknowledge the poll.

Each frame uses a number to identify the present frame and the next frame expected. These are carried in information frames and acknowledgments are also carried in these frames. If there is no outgoing information frame, a receiver ready (RR) frame can be sent to indicate which frame is expected next. If a frame number that is not expected arrives, then a reject response can be used to indicate which frame should arrive.

Transmission can be stopped with a receiver not ready (RNR) signal, which indicates the frame number that was received last (this acknowledges the receipt of the frames that preceded it) and requests that no more frames be sent. This along with the basic packet of seven frames are the flow control mechanisms. If the system delay and transmission bandwidth allow it, the frame format can be extended up to 127 frames.

A potential disadvantage of HDLC for line control in a network is the artificial imposition of a sequencing relation between units. This constraint can cause additional delays when errors occur.

Flow control generally applies to point-to-point procedures between adjacent nodes, while congestion control is a multipoint problem involving keeping the aggregate network resources from becoming oversubscribed.

A typical congestion-prone system is illustrated in Figure 10-14. The throughput rises with increasing traffic and the delay grows until the throughput reaches a maximum. Now, the system is saturated and increasing traffic causes the throughput to drop rapidly due to congestion. If the throughput is allowed to reach zero under this condition, then system deadlock occurs. This can happen when data to two nodes are routed in opposite directions; then each node becomes filled to capacity with messages destined for the other.

The X.25 standard also implements this layer as one of its three layers. ×.25 is the packet switching standard of CCITT. It contains three layers: physical, data link, and network.

ADDRESSING

At each protocol layer there is the need to identify at least the destination of each unit of data. This destination information is needed, since at all layers except possibly the physical layer there may be more than one potential destination.

The most common approach in a layered architecture is to create a hierarchical address space. Two nodes that are physically connected in a shared local loop will have a unique link address. The interface between the two nodes and the local loop may be one of many in the system so the two nodes must have identifiers. The system may have a number of operating programs or processes. These must have logical input/output channels or ports to operate on and these must be addressed properly if the data are to be sent across the network and through the many protocol layers to the desired destination. If there is more than one network, these will all need specific identifiers for multinet operation.

The assignment of addresses may be static or dynamic. Most networks

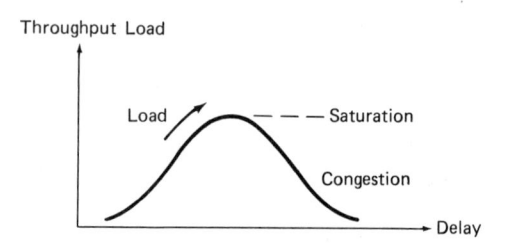

Figure 10-14. A congestion-prone network.

supply two types of addresses, virtual and physical. In a physically addressed network, the address identifies the actual connections between units. In a virtually addressed network, tables are maintained between virtual and physical addresses. This allows units to be connected to more than one physical address while using only one virtual address. Several virtual addresses can also map to the same physical address, allowing several virtual units to be emulated by one physical unit. Those frames whose source address is recognized by the interface are passed to the network layer.

The National Bureau of Standards (NBS) has established the Transport Control Protocol (TCP) and the Internet Protocol (IP). Several other protocols have come to be accepted as de facto standards for telecommunication applications. These include the File Transfer Protocol (FTP), the Virtual Terminal Protocol (VTP), and the Mail Transfer Protocol (MTP).

In most cases, the user will add to the header or trailer fields of a frame the additional protocol information necessary to ensure the proper application of the terminal equipment and the proper processing of the message or data field of the frame. Exactly what must be added depends on the different pieces of machines, subsystems, equipment, and instruments that make up a particular automation system implementation.

SOFTWARE PORTABILITY

Because of the considerable time required to fully develop application software and because requirements for specific and unusual peripheral equipment may be difficult to forecast, it is cost effective for such software to be fully portable, that is, for it to be usable on the widest possible range of equipment.

If proper attention is given to the selection of the application software, it is possible to greatly increase the portability of the software so the software can be used in a great variety of applications. The following characteristics are important to portability:

1. The standardization of all functions.
2. A wide selection of ranges.
3. Adjustable addressing in the network.
4. A standardized instruction set.
5. A standardized protocol.

In making the selection, it is also desirable to choose a combination supported by at least two or more suppliers. Of the above characteristics the first three are normally found to some extent in most application packages.

A disadvantage of the one-vendor approach is that the options are limited by the selection of systems and equipment available from the chosen vendor.

Some companies address only one aspect of the entire factory environment. However, they deal with that aspect so well that many customers choose to integrate these products into an overall factory system.

Many manufacturers are special versions of standardized languages and protocols. These versions may consist of the standard version and the manufacturer's additions or they may be a subset of the standardized version. A special version of a language might not be able to use a standardized compiler. For automatic test equipment the standard language has been ATLAS and is now ADA. Other languages such as BASIC, FORTRAN, PL/I, and Pascal may use extensions or enhancements that may be standardized by the Purdue Industrial Control Workshop (PICW) or the Instrumentation Society of America (ISA). Only the extensions to BASIC have been used for most instrumentation. Many popular extensions are now in the process of being standardized. In the case of a protocol and network for instrumentation, the standards most frequently used are IEEE 488 and IEEE 802. Others that are rapidly becoming standardized are PROWAY, which is a version of IEEE 802, and ETHERNET.

An automation system using one of these languages and protocols is bound to have some portability. This allows the hardware and software to be utilized for a variety of purposes in the factory. If another instrument is required to replace a unit being repaired or to balance changes in the production line, another unit from elsewhere in the factory can easily be used.

An extension of IEEE 802 could be used for a network of test systems. ETHERNET and IEEE 802 can interchange data at 10 Mbits/s, which is 500 times faster than RS-232 or RS-449, even at the higher data rate of 9600 baud. The latter techniques would only be used at the lowest levels.

A wide variation in machine architectures can make the sharing of information difficult, requiring the development of hardware and software compatibility at the level of the network. In order to integrate the various functions into a manufacturing environment, it is necessary to consider a number of factors. In extremely high-electrical-noise environments, low-noise or fiber optic cable should be used.

The word size should be at least 16 bits but processing may be in 32- or 64-bit blocks depending on the efficiency needed. The character set should be ASCII for character compatibility, while graphics support should use the IGES standard. The best choice for a language might be ADA because of the real-time multitasking environment.

The protocol should be compatible with the seven-layer OSI structure. HDLC can be used to support the data link layer. IEEE-802- or PROWAY-based techniques can be used to support the lower layers and define the header, data, and trailer fields of the message frame. Additional protocol as required can be implemented with FTPV and VTP rules.

Although the access method can use a combined form of CSMA/CD and token passing by switching between these methods, this requires additional overhead and may not be cost effective. The method used depends on the efficiency required and the expected channel traffic. The cost of implementing these techniques increases as the number of nodes in the network is increased.

MAP

The goal of making equipment compatible with OSI is commendable, but the rate of progress dissatisfies many users who need to connect diverse devices together today. This has resulted in the development of related OSI networks such as the Manufacturing Automation Protocol (MAP) and the Technical and Office Protocol (TOP). General Motors and Boeing have been involved in the development of these networks since 1980.

MAP is a broad-band token-bus network designed for the CIM environment. This requires linking the various islands of information in the factory system:

1. mainframe computer running materials requirements planning (MRP)
2. supermini or microcomputers running computer-aided design
3. real-time mini or microcomputers doing quality control
4. programmable controllers (PCs) scheduling and dispatching parts on the manufacturing line
5. robots doing spot welding and painting
6. numerical control (NC) machines drilling holes and milling parts

MAP offers gains in productivity that only complete integration can allow. This has led GM to make a commitment to installing MAP in all of its plants.

The initial version of MAP can be expected to undergo some major changes. In the application layer, the protocol which was called the Manufacturing Message Format Standard (MMFS) is to be superseded by Electronics Industries Association standard EIA 13'93, which will be known as RS-511. As a result, equipment for the implementation of MAP may bypass MMFS as often as possible. To quickly upgrade MAP in a way that is as inexpensive as possible for users requires that the changes affect the software. Several configurations are possible: (1) Implement layers three through seven in software with layers one and two and an HDLC interface on a token interface module (TIM) which communicates under $\times.25$. (2) Implement layers five through seven in software with a MAP interface board to handle layers one through four.

For manufacturing and process control users, these are rapidly changing times. Charting a course through manufacturing's information islands can be hazardous. Many users have only recently weathered the problems of implementing MRP; now true CIM will require an even greater commitment.

EXERCISES

1. Many factory information models consist of the application data structure plus a collection of application program procedures that define the structure. Discuss the interrelation between the model, the application program, and the manufacturing system.
2. Factory information models can describe parts or systems with inherent geometries and lend themselves naturally to graphical representation. Show how they can have a hierarchical structure which is built using a bottom-up construction process.
3. In manufacturing, a common hierarchy is the assembly-subassembly relationship. It is common to use standard components as basic building blocks. These are often drawn using templates of standard symbolic shapes. Flowcharts and digital logic symbols for 2D models are common examples. The standard components can be defined in their own coordinates and may have not only geometrical data but also associated application data. Discuss how the factory model can show not only which components are present, but also how they are connected using a hierarchy that is created for a variety of purposes.
4. Describe the application of structured design to Exercise 3.
5. Draw a block diagram for a control system for a steam-heated distillation column.
6. What are the advantages and disadvantages of centralized process control?
7. Diagram a remote multiplexing configuration.
8. How can control problems be avoided at higher data rates in position control systems?
9. Diagram the following closed-loop position control techniques: (a) linear velocity ramp, (b) exponential velocity output, (c) programmed deceleration.
10. Discuss the differences between manual and off-line robot programming.
11. How would loop and unit distributed control schemes be used for the optimum results in an NC facility for engine blocks? Describe several benchmarks for evaluating software for this facility.
12. Discuss the characteristics of the following types of networks: (a) master-slave, (b) hierarchical, (c) star, (d) multidrop, (e) ring, (f) peer, (g) point-to-point.
13. Draw a layered network based on the CCITT-ISO standard. At what levels would the following bus standards be used: (a) RS-232C, (b) IEEE 488, (c) IEEE 802/PROWAY, (d) ETHERNET?
14. Discuss the differences between CSMA/CD and token passing.
15. What is network congestion and how can it be avoided?
16. Discuss what is meant by software portability and how it can be achieved.

BIBLIOGRAPHY

Abramson, N. and Kuo, F. F., Eds. *Computer-Communications Networks*. Englewood Cliffs, NJ: Prentice-Hall, 1973.
Adams, L. F. *Engineering Measurements and Instrumentation*. London: English Universities, 1975.
Allison, D. R. A design philosophy for microcomputer architectures. *Computer,* **10,** 2 (1977).
Altman, L., Ed. *Microprocessors*. New York: Electronic Magazine Book Series, 1975.
Anderson, D. A. Design of self-checking digital networks using code techniques. Ph.D. Thesis, Report R 527, University of Illinois, Urbana, IL, 1971.
Andrews, M. *Principles of Firmware Engineering in Microprogram Control*. London: Computer Science, 1980.
Athans, M., and Falb, P., *Optimal Control: An Introduction to the Theory and Its Applications*. New York: McGraw-Hill, 1966.
Avizienis, A. Fault-tolerant computing—An overview. *Computer* (January-February 1971).
Bartogiak, G., Guide to thermocouples. *Instruments and Control Systems* (November 1978).
Bates, R. G., *Determination of pH—Theory and Practice*. New York: Wiley, 1964.
Baum, A., and Senzig, D. Hardware considerations in a microcomputer multiprocessor system. San Francisco: COMCON paper, February 1975.
Bell, C. and Newell, A. *Computer Structures*. New York: McGraw-Hill, 1970.
Benedict, R. P. *Fundamentals of Temperature, Pressure and Flow Measurements*. New York: Wiley, 1969.
Beveridge, G. S. G., and Schechter, R. S. *Optimization: Theory and Practice*. New York: McGraw-Hill Series in Chemical Engineering, 1970.
Blasso, L. Flow measurement under any conditions. *Instruments and Control Systems* (February 1975).
Breuer, M. A., and Griedman, A. D. *Diagnosis and Reliable Design of Digital Systems*. Woodland Hills, CA: Computer Science, 1976.
Brooks, F. P. An overview of microcomputer architecture and software. In *Micro Architecture*, EUROMICRO 1976 Proceedings.
Brooks, F. P. *The Mythical Man-Month, Essays on Software Engineering*. Reading, MA: Addison-Wesley, 1975.
Buckley, P. S., *Techniques of Process Control*. New York: Wiley, 1964.
Burton, D. P. Handle microcomputer I/O efficiently. *Electronic Design* 13 (June 21, 1978).
Burzio, G., Operating systems enhance μCs. *Electronic Design* (June 21, 1978).
Butler, J. N., *Ionic Equilibrium: A Mathematical Approach*. Reading, MA: Addison-Wesley, 1964.
Caldwell, W. I., Coon, G. A., and Zoss, L. M. *Frequency Response for Process Control*. New York: McGraw-Hill, 1959.
Camenzind, H. R. *Electronic Integrated System Design*. New York: Van Nostrand Reinhold, 1972.
Chandy, K. M., and Reiser, M., Eds. *Computer Performance*. Amsterdam; North-Holland, 1977.
Childs, R. E. Multiple microprocessor systems: Goals, limitations and alternatives. In *Digest of Papers, COMPCON, Spring 79*, 1979.
Chu, Y. *Computer Organization and Microprogramming*. Englewood Cliffs, NJ: Prentice-Hall, 1972.
Coffee, M. B. Common-mode rejection techniques for low-level data acquisition. *Instrumentation Technology* (July 1977).
Cohen, T. Structured flowcharts for multiprocessing. *Computer Languages,* 13 (1978).

286 BIBLIOGRAPHY

Combs, C. F., Ed. *Basic Electronic Instrument Handbook.* New York: McGraw-Hill, 1972.

Considine, D. M. *Process Instruments and Controls Handbook.* New York, McGraw-Hill, 1957.

Crick, A. Scheduling and controlling I/O operations. *Data Processing* (May-June 1974).

Cutler, H. Linear velocity ramp speeds stepper and servo positioning. *Control Engineering.* **24,** 5 (1977).

Dal Cin, M. Performance evaluation of self-diagnosing multiprocessor systems. Presented at Conference on Fault-Tolerant Computing, Toulouse, France, 1978.

Daniels, F. *Outlines of Physical Chemistry.* New York: Wiley, 1948.

Diefenderfer, A. J. *Principles of Electronic Instrumentation.* Philadelphia: Saunders, 1972.

Dijkstra, E. W. *A Discipline of Programming.* Englewood Cliffs, NJ: Prentice-Hall, 1976.

Doebelin, E. O. *Measurement System—Application and Design.* New York: McGraw-Hill, 1975.

Doherty, D. W., and Wells, E. J. Digital power drive dynamics are characterized by ROMs. *Control Engineering,* **25,** 1 (1978).

Donovan, J. J. *Systems Programming.* New York: McGraw-Hill, 1972.

Dowsing, R. D. Processor management in a multiprocessor system. *Electronic Letters* (November 1976).

Drucker, P. F. *The Age of Discontinuity: Guidelines to Our Changing Society.* New York: Harper & Row, 1968.

Dube, R., Herron, G. J., Little, F. F., and Riesenfeld, R. F. SURFED: An interactive editor for free-form surfaces. *Computer Aided Design,* **10,** 2 March (1978).

Eastman, C., Lividini, J., and Stoker, D. A database for designing large physical systems, AFIPS, NCC 1975.

Eastman, C. M., and Henrion, M. GLIDE: a language for design information systems, In SIGGRAPH '77 Proceedings, published as *Computer Graphics,* **11,** 2 (1977).

Eckhouse, R. H., Jr. *Minicomputer Systems.* Englewood Cliffs, NJ: Prentice-Hall, 1975.

Eckman, D. P. *Automatic Process Control.* New York: Wiley, 1958.

E. E. U. A., *Installation of Instrumentation and Process Control Systems,* Handbook No. 34. London: Constable, 1973.

Elliott, T. C. Temperature, pressure, level, flow-key measurements in power and process. *Power* (September 1975).

Encarnacao, J., et al. The workstation concept of GKS and the resulting conceptual differences to the GSPC core system. In SIGGRAPH '80 Proceedings, published as *Computer Graphics,* **14,** 3 (July 1980).

Engel, S., and Granda, R. Guidelines for man/display interfaces. Technical Report TR 00.2720. Poughkeepsie, NY: IBM, 1975.

English, R. E. Systems management of a CAD/CAM installation. In *CADCON East 84,* Boston, 1984. New York: Morgan-Grampian, 1985.

Enslow, P. H., Jr., Ed. *Multiprocessors and Parallel Processing.* New York: Wiley, 1974.

Estes, V. Robots—The key to the factory with a future. Presented at Automated Manufacturing 1984, Greenville, SC.

Evans, F. L, Jr. *Equipment Design Handbook for Refineries and Chemical Plants.* Houston: Gulf, 1971.

Fablrycky, W. J., Ghare, P. M., and Torgersen, P. E. *Industrial Operations Research.* Englewood Cliffs, NJ: Prentice-Hall, 1972.

Farnbach, W. A. Bring up your μP bit-by-bit. *Electronic Design,* **24,** 15 (July 1976).

Faux, I. D., and Pratt, M. J. *Computational Geometry for Design Manufacture,* New York: Wiley, 1979.

Fegreus, Jack. The mapping of open communications. *Digital Review* (February 1986).

Feiner, S., Nagy, S., and van Dam, A. An integrated system for creating and presenting complex computer-based documents. In Proceedings 1981 SIGGRAPH Conference, published as *Computer Graphics,* **15,** 3 (August 1981).

Fenves, S. J., and Branin, F. H., Jr. A network-topological formulation of structural analysis, *ASCE Journal of Structural Division,* (August 1963).

Fields, A., Maisano, R., and Marshall, C. A comparative analysis of methods for tactical data inputting. Army Research Institute, 1977.

Finkel, J. I. Improved software for IC layout. *Computer Aided Engineering* (November/December 1983).

Foley, J. D. Evaluation of small computers and display controls for computer graphics. *Computer Group News,* **3,** 1 (January/February 1970).

Foley, J. D. Managing the design of user-computer interfaces. *Computer Graphics World* (December 1983).

Foley, J. D., and Wallace, V. L. The art of natural graphic man-machine conversation. *Proceedings of the IEEE,* **62**(4) (April 1974).

Foley, J. D., Wallace, V., and Chan, P. The human factors of interaction techniques. George Washington University Institute for Information Science and Technology Technical Report GWU-IIST-81-03, Washington, DC, 1981.

Forrest, A. R. Mathematical principles for curve and surface representation in Curved Surfaces in Engineering, I. J. Brown, Ed., IPC Science and Technology Press Ltd., Guildford, Surrey, England, 1972.

Forrest, D. R. On coons and other methods for the representation of curved surfaces. *Computer Graphics and Image Processing,* (1) 4 (December 1972).

Foskett, R. Torque measuring transducers. *Instruments and Control Systems* (November 1968).

Foster, C. C. *Computer Architecture.* New York: Van Nostrand Reinhold, 1970.

Franklin, M. A., Kahn, S. A., and Stucki, M. J., Design issues in the development of a modular multiprocessor communications network. Presented at Sixth Annual Symposium on Computer Architecture, 1979.

Freedman, M. D. Principles of digital computer operation. New York: Wiley, 1972.

Friedman, A. D., and Memon, P. R. Fault detection in digital circuits. Englewood Cliffs, NJ: Prentice-Hall, 1971.

Fuchs, H. Distributing a visible surface algorithm over multiple processors. In Proceedings, 1977 ACM National Conference.

Fuchs, H., Duran, J., and Johnson, B. A system for automatic acquisition of three-dimensional data. In *Proceedings of the 1977 NCC,* AFIPS Press.

Fuchs, H., Kedem, Z. M., and Uselton, S. P. Optimal surface reconstruction from planar contours. *Communications of the ACM,* **20,** 19 (October 1977).

Fung, K. T., and Torng, H. C. On the analysis of memory conflicts and bus contentions in a multiple-microprocessor system. *IEEE Transactions on Computers,* **C-27,** 1 (January 1979).

Fung, K. T., and Torng, H. C. On the analysis of memory conflicts and bus contentions in a multiple-microprocessor system. *IEEE Transactions on Computers,* **C-27,** 1 (January 1979).

Garland, H. *Introduction to Microprocessor System Design.* New York: McGraw-Hill, 1979.

Garrett, M. A. Unified non-procedural environment for designing and implementing graphical interfaces to relational data base management systems, Ph.D. Thesis, The George Washington University, 1980.

Gear, C. W. *Computer Organization and Programming.* New York: McGraw-Hill, 1974.

Germann, J. J. Using special processors to enhance engineering workstations. In *CAD/CAM West 84,* San Francisco, 1984. Morgan-Grampian: New York, 1984.

Geyer, K. E., and Wilson, K. R. Computing with feeling. In Proceedings, IEEE Conference on Computer Graphics, Pattern Recognition and Data Structure, 1975.

288 BIBLIOGRAPHY

Gibson, J. E. *Nonlinear Automatic Control*. New York: McGraw-Hill, 1963.

Giloi, W. *Interactive Computer Graphics—Data Structures, Algorithms, Languages*. Englewood Cliffs, NJ: Prentice-Hall, 1978.

Goldberg, A., and Robson, D. A metaphor for user interface design. Palto Alto, CA: Xerox Palo Alto Research Center Publication, 1979.

Gonzalez, M. J., and Ramamoorthy, C. V. Parallel task execution in a decentralized system. *IEEE Transactions on Computers* (December 1972).

Gordon, W. J., and Riesenfeld, R. F. Bernstein-Bezier methods for the computer-aided design of free-form curves and surfaces. *Journal of the ACM*, **21**, 2 (April 1974).

Grant, E., and Leavenworth, R. S. *Statistical Quality Control*. New York: McGraw-Hill, 1972.

Gregory, B. A. *An Introduction to Electrical Instrumentation*. New York: Macmillan, Inc., 1973.

Griffiths, J. G. A surface display algorithm. *Computer Aided Design*, **10**, 1 (January 1978).

Grimsdale, R. L., and Johnson, D. M. A modular executive for multiprocessor systems. Sheffield, England: Trends in On-Line Computer Control Systems, Conference 1972.

Groff, G. K., and Muth, I. F. *Operations Management: Analysis for Decisions*. Homewood IL: Irwin, 1972.

Guedj, R., et al., Ed. *Methodology of Interaction*, North-Holland, Amsterdam, 1980.

Hall, J. Flowmeters—Matching applications and devices. *Instruments and Control Systems* (February 1978).

Hall, J. Solving tough flow monitoring problems. *Instruments and Control Systems* (February 1980).

Hamilton, M., and Zeldin, S. Higher order software—A methodology for defining software. *IEEE Transactions on Software Engineering*, **SE-2**, 1 (1976).

Hanau, P. R., and Lenorovitz, D. R. Prototyping and simulation tools for user/computer dialogue design. In SIGGRAPH '80 Proceedings, published as *Computer Graphics*, **14**, 2 (July 1980).

Hansen, W. User engineering principles for interactive systems. In Proceedings, 1971 Fall Joint Computer Conference.

Harriott, P. *Process Control*. New York: McGraw-Hill, 1964.

Harris, J. A., and Smith, D. R. Hierarchical multiprocessor organizations. In Proceedings, Fourth Annual Symposium on Computer Architecture, 1977.

Hayes, P., Ball, E., and Reddy, R. Breaking the man-machine communication barrier. *Computer*, **14**, 3 (March 1981).

Helmers, C. T., Ed. *Robotics Age, in the Beginning*. Rochelle Park, NJ: Hayden, 1983.

Herot, C. F., et al. A prototype spatial data base management system. In SIGGRAPH '80 Proceedings, published as *Computer Graphics*, **14**, 2 (July 1980).

Herrick, C. N. *Instrumentation and Measurement for Electronics*. New York: McGraw-Hill, 1972.

Hill, F. J., and Peterson, G. R., *Digital Systems: Hardware Organization and Design*. New York: Wiley, 1973.

Hnatek, Eugene R. *A User's Handbook of Semiconductor Memories*. New York: Wiley, 1977.

Hnatek, Eugene R. Current semiconductor memories, *Computer Design* (April 1978).

Hodges, D. A. *Semiconductor Memories*. New York: IEEE, 1972.

Hordeski, M. F. Digital control of microprocessors. *Electronic Design* (December 1975).

Hordeski, M. F. Digital sensors simplify digital measurements. *Measurements and Data* (May-June 1976).

Hordeski, M. F. When should you use pneumatics, when electronics? *Instruments and Control Systems* (November 1976).

Hordeski, M. F. Guide to digital instrumentation for temperature, pressure instruments. *Oil, Gas and Petrochem Equipment* (November 1976).

Hordeski, M. F. Digital instrumentation for pressure, temperature/pressure, readout instruments, *Oil, Gas and Petrochem Equipment* (December 1976).

Hordeski, M. F. Innovative design: Microprocessors. *Digital Design* (December 1976).

Hordeski, M. F. Passive sensors for temperature measurement. *Instrumentation Technology* (February 1977).

Hordeski, M. F. Adapting electric actuators to digital control. *Instrumentation Technology* (March 1977).

Hordeski, M. F. Fundamentals of digital control loops and factors in choosing pneumatic or electronic instruments. Presented at the SCMA Instrumentation Short Course, Los Angeles, 1977.

Hordeski, M. F. Balancing microprocessor-interface tradeoffs. *Digital Design* (April 1977).

Hordeski, M. F. Digital position encoders for linear applications. *Measurements and Control* (July-August 1977).

Hordeski, M. F. Future microprocessor software. *Digital Design* (August 1977).

Hordeski, M. F. Radiation and stored data. *Digital Design* (September 1977).

Hordeski, M. F. Microprocessor chips. *Instrumentation Technology* (September 1977).

Hordeski, M. F., Process controls are evolving fast. *Electronic Design* (November 1977).

Hordeski, M. F., Fundamentals of digital control loops. *Measurements and Control* (February 1978).

Hordeski, M. F., Using microprocessors. *Measurements and Control* (June 1978).

Hordeski, M. F., Interfacing microcomputers in Control Systems, Instruments and Control Systems, November 1978.

Hordeski, M. F., *Illustrated Dictionary of Microcomputer Terminology*. Blue Ridge Summit, PA: Tab, 1978.

Hordeski, M. F., *Microprocessor Cookbook*. Blue Ridge Summit, PA: Tab, 1979.

Hordeski, M. F., Selecting test strategies for microprocessor systems. In *ATE Seminar Proceedings*, Pasadena, CA, 1982. New York: Morgan-Grampian, 1982.

Hordeski, M. F., Selection of a test strategy for MPU systems. *Electronics Test* (February 1982).

Hordeski, M. F., Trends in displacement sensors. In *Sensors and Systems Conference Proceedings*, Pasadena, CA, 1982 (Network Exhibitions, Campbell, CA).

Hordeski, M. F., The impact of 16-bit microprocessors. In *Instrumentation Symposium Proceedings*, Las Vegas, 1982. Research Triangle Park, NC: Instrument Society of America, 1982.

Hordeski, M. F., Diagnostic strategies for microprocessor systems. In *ATE Seminar Proceedings*, Anaheim, CA, 1983. New York: Morgan-Grampian, 1983.

Hordeski, M. F., *Microprocessors in Industry*. New York: Van Nostrand Reinhold, 1984.

Hordeski, M. F., *The Design of Microprocessor Sensor and Control Systems*. Reston, VA.: Reston, 1984.

Hordeski, M. F., CAD/CAM equipment reliability. Presented at Western Design Engineering Show and ASME Conference, San Francisco, 1984.

Hordeski, M. F., Specifying and selecting CAD/CAM equipment. In *CADCON West*, Anaheim, CA, 1985. New York: Morgan-Grampian, 1985.

Hordeski, M. F., A tutorial on CIM/Factory Automation, Presented at Western Design Engineering Show and ASME Conference, Anaheim, CA, 1985.

Hordeski, M. F., *CAD/CAM Techniques*. Reston, VA.: Reston, 1986.

Hordeski, M. F., *Microcomputer Design*. Reston, VA.: Reston, 1986.

Hornbuckle, G. D., The computer graphics/user interface. *IEEE Transactions* **HFE-8** (1) (March 1967).

Hougen, J. O., *Measurements and Control Applications*. Research Triangle Park, NC: Instrument Society of America, 1979.

Houtzel, A., The graphics side of group technology. In *CADCON East 84*, Boston, 1984. New York: Morgan-Grampian, 1985.

Hubschman, H., and Zucker, S., Frame-to-frame coherence and the hidden surface computation: Constraints for a convex world. In SIGGRAPH '81 Proceedings, published as *Computer Graphics*, **15,** 3 (August 1981).

290 BIBLIOGRAPHY

Hudry, J., Man-machine interface issues. *Computer Graphics World* (April 1984)
Husson, S. S., *Microprogramming: Principles and Practices*. Englewood Cliffs, NJ: Prentice-Hall, 1970.
Hutchison, J. W., *ISA Handbook of Control Valves*. Pittsburgh: Instrument Society of America, 1971.
Intel Corp. 8086 User's Guide. Santa Clara, CA: Intel, 1976.
Irani, K., and Wallace, V., On network linguistics and the conversational design of queueing networks. *Journal of the ACM*, **18** (October 1971).
Jackson, M. A., *Principles of Program Design*. New York: Academic, 1975.
Janki, C., What's new in motors and motor controls? *Instruments and Control Systems* (November 1979).
Jensen, K., and Wirth, N., *Pascal User Manual and Report*, 2nd, ed. New York: Springer-Verlag, 1974.
Johnson, C., Solids modeling. In *CADCON East 84*, Boston, 1984. New York: Morgan-Grampian, 1985.
Johnson, S., and Lesk, M., Language development tools. *The Bell System Technical Journal*, **57** (6, 2) (1978).
Jones, B. E., *Instrumentation, Measurement and Feedback*. New York: McGraw-Hill, 1977.
Jones, E. B., *Instrument Technology*. Woburn, MA: Butterworth, Vol. 1 (revised), 1965; Vol. 2, 1956.
Jones, J. C., *Design Methods*. New York: Wiley-Interscience, 1970.
Jordan, B. W., Lennon, W. J., and Holm, B. C., An improved algorithm for the generation of non-parametric curves. *IEEE Transactions on Computers*, **C-22** (12) (December 1973).
Jutila, J. M., Temperature instrumentation. *Instrumentation Technology* (February 1980).
Kaplan, M., and Greenberg, D., Parallel processing techniques for hidden surface algorithms. In *SIGGRAPH '79 Proceedings*, published as *Computer Graphics* (August 1979).
Kaufman, A. B., Monitor acceleration, velocity or displacement. *Instruments and Control Systems* (October 1979).
Kay, A. C., Microelectronics and the personal computer. *Scientific American*, **237**, 3 (September 1977).
Kennedy, J. R., *A System for Timesharing Graphic Consoles, FJCC*. Washington, DC: Sparton, 1966.
Kilgour, A. C., The evolution of a graphic system for linked computers. *Software—Practice and Experience*, **1** (1971).
Klingman, E. E., *Microprocessor Systems Design*. Englewood Cliffs, NJ: Prentice-Hall, 1977.
Knapp, J. M., The ergonomic millennium. *Computer Graphics World* (June 1983).
Klipec, B., How to avoid noise pickup on wire and cables. *Instruments and Control Systems* (December 1977).
Knuth, D. E., *The Art of Computer Programming. Volume 1: Fundamental Algorithms*. Reading, MA: Addison-Wesley, 1973.
Kohonen, T., *Digital Circuits and Devices*. Englewood Cliffs, NJ: Prentice-Hall, 1972.
Kolk, W. R., PRM—Control candidate in energy limited systems. *Control Engineering*, **24**, 10 (1977).
Kriloff, H., Human factor considerations for interactive display systems. In S. Trem, Ed., *Proceedings, ACM/SIGGRAPH Workshop on User-Oriented Design of Interactive Graphics Systems*, ACM, 1976.
Kuck, D. J., *The Structure of Computers and Computations*, Vol. 1. New York: Wiley, 1978.
Kuenning, M. K., Programmable controllers: Configuration and programming. Presented at Automated Manufacturing 1984, Greenville, SC.

Kwakernaak, H., and Swan, R., *Linear Optimal Control Systems*. New York: Wiley, 1972.
Lampson, W., Bravo manual. In *Alto User's Handbook*. Palo Alto, CA: Xerox Palo Alto Research Center, 1978.
Lane, J., and Carpenter, L., A generalized scan line algorithm for the computer display of parametrically defined surfaces. *Computer Graphics and Image Processing*, **11**, 1979.
Lawrence, S., and Marcus, L. S., Designing PC boards with a centralized database. *Computer Graphics World* (March 1984).
Leahy, W., Data base management–Automated graphics generation. In *CADCON East 84*, Boston, 1984. New York: Morgan-Grampian, 1985.
Leininger, M., Present and future developments in robotic applications. Presented at Automated Manufacturing 1984, Greenville, SC.
Levenspiel, O., *Chemical Reaction Engineering*. New York: Wiley, 1962.
Leventhal, L. V., *Microprocessor: Software, Hardware, Programming*. Englewood Cliffs, NJ: Prentice-Hall, 1978.
Liptak, B. G., *Environmental Engineers' Handbook*, Vols. I–III. Radnor, PA: Chilton, 1974.
Liptak, B. G., *Instrumentation in the Processing Industries*. Radnor, PA: Chilton, 1973.
Liptak, B. G., *Instrument Engineers' Handbook*. Radnor, PA: Chilton Book Co., Vol. I, 1969; Vol. II, 1970; Supplement, 1972.
Liptak, B. G., Ed., *Instrument Engineers' Handbook on Process Measurement*. Radnor, PA: Chilton, 1980.
Liptak, B. G., Ultrasonic instruments. *Instrumentation Technology* (September 1974).
Lloyd, G., Managing the CAD transition in design. Presented at *CADCON East 84*, Boston, 1984. New York: Morgan-Grampian, 1985.
Lomas, D. J., Selecting the right flowmeter. *Instrumentation Technology* (1977).
Lorin, H., *Parallelism in Hardware and Software*. Englewood Cliffs, NJ: Prentice-Hall, 1972.
Lunden, J. W., Intelligent vision: Rapid advances in industrial automation. Presented at Automated Manufacturing 1984, Greenville, SC.
Lupfer, D. E., and Johnson, M. L., Automatic control of distillation columns to achieve optimum operation. In *ISA Transactions*. Pittsburgh: Instrument Society of America, 1974.
Madnick, S. F., and Donovan, J. L., *Operating Systems*. New York: McGraw-Hill, 1974.
Magnenat-Thalmann, N., and Thalmann, D., A graphical Pascal extension based on graphical types. *Software—Practice and Experience*, **11** (1981).
Martin, D. P., *Microcomputer Design*. Chicago: Martin Research, 1975.
Martin, J., *Design of Man-Computer Dialogues*. Englewood Cliffs, NJ: Prentice-Hall, 1973.
Martin, W. A., Computer input/output of mathematical expressions. 2nd Symposium on Symbolic Algebraic Manipulation, ACM, March 1971.
Mazur, T., Microprocessor basics: Part 4: the Motorola 6800. *Electronic Design* (July 1976).
McCool, M., Interfacing CAD/CAM with ATE. In ATE West Proceedings, 1983.
McGlynn, D. R., *Microprocessors*. New York: Wiley, 1976.
McKay, C. W., An approach to distributing intelligence among a network of cooperative, autonomous, functional computing clusters. Presented at Automated Manufacturing 1984, Greenville, SC.
McKenezie, K., and Nichols, A. J., Build a compact microcomputer. *Electronic Design*, **24**, 10 (1976).
McLain, D. H., Computer construction of surfaces through arbitrary points, IFIP. Amsterdam: North-Holland, 1974.
McManigal, D., and Stevenson, D., Architecture of the IBM 3277 graphics attachment. *IBM Systems Journal*, **19**, 3 (1980).
Meditch, J. S., Stochastic optimal linear estimation and control. New York: McGraw-Hill, 1969.

BIBLIOGRAPHY

Medlock, R. S., Vortex shedding meters. Presented at Liquefied Gas Symposium, London, 1978.

Metcalfe, R. M., and Boggs, D. R., ETHERNET: Distributed packet switching for local computer networks. *Communications of the ACM*, **19**, 7 (1976).

Meyer, J. D., Commercial machine vision systems. *Computer Graphics World* (October 1983).

Meyrowitz, N., and Moser, M., BRUWIN: An adaptable design strategy for window manager/virtual terminal systems. In Proceedings of the 8th Annual Symposium on Operating Systems Principles (SIGOPS), Pacific Grove, CA, 1981.

Miller, N., Bus-oriented graphics systems. *Computer Graphics World* (May 1983).

Miller, R. B., Response time in man-computer conversational transactions. In *1968 FJCC, AFIPS Conference Proceedings*, Vol. 33. Montvale, NJ: AFIPS, 1968.

Minardi, L. R., CAD/CAM's drawing/model dichotomy. *Computer Graphics World* (January 1983).

Moran, T., The command language grammar: A representation for the user interface of interactive computer systems. *International Journal of Man-Machine Studies*, **15** (1981).

Moss, D., Multiprocessing adds muscle to μPs. *Electronic Design*, **II** (1978).

Motorola Semiconductor. *MC6800 Microprocessor User's Manual*. Austin, TX: Motorola, 1979.

Motorola Semiconductor. *M6800 Microprocessor Applications Manual*. Phoenix, AZ: Motorola, 1975.

Mullins, M., Instrument controllers—Evaluating cost and function. *Test & Measurement World* (April 1984).

Murphy, H. N., Flow measurement by insertion turbine meters. Presented at Instrument Society of America Symposium, Wilmington, Delaware, 1979.

Murrill, P. W., *Automatic Control of Processes*. Scranton, PA: International Textbook, 1967.

Myers, G. J., *Reliable Software through Composite Design*. New York: Petrocelli Charter, 1975.

Myers, G. J., *Advances in Computer Architecture*. New York: Wiley, 1978.

Newman, W. M., and Sproull, R. F., *Principles of Interactive Computer Graphics*, 2nd ed. New York: McGraw-Hill, 1979.

Newton, R. S., An exercise in multiprocessor operating system design. Presented at the Agard Conference on Real-Time Computer-based Systems, Athens, 1974.

Ng, N., and Marsland, T., Introducing graphics capabilities to several high-level languages. *Software—Practice and Experience*, **8** (1978).

Nick, J. R., Using Schottky 3-state outputs in bus-organized systems. *Electronic Design News*, **19**, 23 (1974).

Norton, F. J., CADD, human relations, and the management process. In *CADCON West 84*, San Francisco, 1984. New York: Morgan-Grampian, 1984.

Norton, H. N., *Handbook of Transducers for Electronic Measuring Systems*. Englewood Cliffs, NJ: Prentice-Hall, 1969.

Novitsky, M. P., MRP in the process industry. Presented at Automated Manufacturing 1984, Greenville, SC.

Nunn, M., CAE/CAD/CAM testability—An overview. In *CADCON West 84*, San Francisco, 1984. New York: Morgan-Grampian, 1984.

O'Brien, Michael T., A network graphical conferencing system. Santa Monica, CA: Rand Corp., 1979 (Report No. N-1250-DARPA).

Oliver, B. M. and Cage, J. M., *Electronic Measurements and Instrumentation*. New York: McGraw-Hill, 1971.

Olmstead, K., The future factory—A first report. *Test & Measurement World* (December 1983).

Oppenheim, A. V., and Schafer, S., *Digital Signal Processing*. Englewood Cliffs, NJ: Prentice-Hall, 1975.

Ottinger, L., Using robots in flexible manufacturing cells/facilities. Presented at Automated Manufacturing 1984, Greenville, SC.

Palmer, R., Nonlinear feedforward can reduce servo settling time. *Control Engineering*, **26**, 3 (1978).

Park, R. M., Applying the systems concept to thermocouples. *Instrumentation Technology* (August 1973).

Parsons, W. A., *Chemical Treatment of Sewage and Industrial Wastes*. Washington, DC: National Lime Association, 1965.

Patel, J. H., Processor-memory interconnections for multiprocessors. In Proceedings, Sixth Annual Symposium on Computer Architecture, 1979.

Patterson, D. A., and Seguin, C. H., Design considerations for single-chip computers of the future. *IEEE Transactions on Computers*, **C-29** (February 1980).

Pearson, D. J., Graphics workstation intelligence. *Computer Graphics World* (January 1983).

Perry, J. H., *Chemical Engineers' Handbook*, 4th ed. New York: McGraw-Hill, 1963.

Peters, G. J., Interactive computer graphics application of the bi-cubic parametric surface to engineering design problems. NCC. Montvale, NJ: AFIPS, 1974.

Peters, L. J., and Tripp, L. L., Is software design wicked? *Datamation*, **22**, 6 (1976).

Peuto, B. L., and Shustek, L. J., Current issues in the architecture of microprocessors. *Computer*, (February 1977).

Pferd, W., A new boost with automated data capture. *Computer Graphics World* (March 1983).

Piller, E., Real-time scan unit with improved picture quality. *Computer Graphics*, **14** (1 & 2) (July 1980).

Pinto, J., Artificial intelligence and robotics in the future factory. *Test & Measurement World* (December 1983).

Plumb, H. H., *Temperature: Its Measurement and Control in Science and Industry*. Research Triangle Park, NC: Instrument Society of America, 1972.

Preiss, R., Storage CRT display terminals: Evolution and trends. *Computer*, **11** (11) (November 1978).

Ramot, J., Nonparametric curves. *IEEE Transactions on Computers*, **C-25** (1) (January 1976).

Renfrow, N., Tools for facilities management. In *CADCON East 84*, Boston, 1984. New York: Morgan-Grampian, 1984.

Requicha, A., Representations for rigid solids: Theory, methods, and systems. *Computing Surveys*, **12** (4) (December 1980).

Riesenfeld, R. F., Aspects of modeling in computer-aided geometric design, National Computer Conference Proceedings, p. 597, 1975.

Riesenfeld, R. F., Non-uniform B-spline curves. In Proceedings, 2nd USA-Japan Computer Conference, 1975.

Riley, J., Process control for a PWB facility. Presented at Automated Manufacturing 1984, Greenville, SC.

Roberson, R. E., Automated manufacturing and management decision information. Presented at Automated Manufacturing 1984, Greenville, SC.

Rodgers, R. C., Peripherals for programmable controllers. *Computer Aided Engineering* (November/December 1982).

Ross, C. A., Automation policies, practices and procedures. Presented at Automated Manufacturing 1984, Greenville, SC, 1984.

Roth, G., Integrated test functions aid in process control. *Test & Measurement World* (December 1983).

Rubin, F., Generation of nonparametric curves. *IEEE Transactions on Computers*, **C-25** (1) (January 1976).

Schaeffer, E. J., and Williams, T. J., An analysis of fault detection correction and prevention in industrial computer systems. Purdue University Laboratory for Applied Industrial Control, 1977.

Schmidt, W. C., The promise of automatic digitizing. *Computer Graphics World* (April 1983).
Seybold, J., The Xerox professional workstation. *The Seybold Report*, **10** (16) (April 1981).
Sheingold, D. H., *Analog-Digital Conversion Handbook*. Norwood, MA: Analog Devices, 1972.
Shepherd, M., Jr., Distributed computing power: Opportunities and challenges. Presented at National Computer Conference, 1977.
Shinskey, F. G., *Distillation Control*. New York: McGraw-Hill, 1977.
Shinskey, F. G., Energy conservation through control. New York: Academic, 1978.
Shinskey, F. G., *pH and Plan Control*. New York: Wiley, 1973.
Shinskey, F. G., *Process Control Systems*. New York: McGraw-Hill, 1979.
Shneiderman, B., Human factors experiments in designing interactive systems. *Computer*, **12**, 12 (December 1979).
Sigma Instruments. *Stepping Motor Handbook*. Braintree, MA: Sigma Instruments, 1972.
Sigsbly, B., ATE in the future factory. *Test & Measurement World* (December 1983).
Skrokov, M. R., Ed., *Mini and Microcomputer Control in Industrial Processes*. New York: Van Nostrand Reinhold, 1980.
Slomiana, M., Selecting differential pressure instruments. *Instrumentation Technology* (August 1979).
Smith, C. L., *Digital Computer Process Control*. Scranton, PA: International Textbook, 1972.
Smythe, M., Applications in mechanical CAD. In *CADCON East 84*, Boston, 1984. New York: Morgan-Grampian, 1984.
Sneeringer, J., User-interface design for text editing: A case study. *Software—Practice and Experience*, **8** (1978).
Sobel, H. S., *Introduction to Digital Computer Design*. Reading, MA: Addison-Wesley, 1970.
Socci, V., Microprocessors in distributed graphics. *Computer Graphics World* (May 1983).
Soisson, H. E., *Instrumentation in Industry*. New York: Wiley, 1975.
Soucek, B., *Microprocessors and Microcomputers*. New York: Wiley, 1976.
Spink, L. K., *Principles and Practices of Flowmeter Engineering*, 9th ed. Foxboro, MA: The Foxboro Company, 1967.
Spitzer, F., and Howarth, B., *Principles of Modern Instrumentation*. New York: Holt, Rinehart and Winston, 1972.
Sproull, R. F., and Thomas, E. L., A network graphics protocol. *Computer Graphics*, **8**, 3 (Fall 1974).
Sroczynski, C., 3D modeling in process and power plant design. In *CADCON East 84*, Boston, 1984. New York: Morgan-Grampian, 1984.
Stevens, W. P., Myers, G. J., and Constantine, L. L., *Structural Design*. New York: Yourdon, 1975.
Stone, Harold, Critical load factors in two-processor distributed systems. *IEEE Transactions on Software Engineering*, **SE-4** (3) (May 1978).
Stone, H. S., *Introduction to Computer Architecture*. New York: McGraw-Hill, 1975.
Stover, R. N., Automating database capture for CAD/CAM. *Computer Graphics World* (March 1983).
Tagg, C. F., *Electrical Indicating Instruments*. Woburn, MA: Butterworth, 1974.
Tasar, O., and Tasar, V., A study of intermittent faults in digital computers. In *AFIPS Conference Proceedings*, Vol. 46, Montvale, NJ: AFIPS, 1977.
Taylor, A. P., Getting a handle on factory automation. *Computer-Aided Engineering* (May-June 1983).
Technical Data Book—Petroleum Refining. Washington, DC: American Petroleum Institute, 1970.
Texas Instruments. *TMS 9900 Microprocessor Data Manual*. Dallas: Texas Instruments, 1978.
Texas Instruments. *The TTL Data Book for Design Engineers*. Dallas: Texas Instruments, 1973.
Texas Instruments. *The Microprocessor Handbook*. Houston: Texas Instruments, 1975.
Thomas, T. B., and Arbuckle, W. L., Multiprocessor software: two approaches. Presented at Conference on the Use of Digital Computers in Process Control, Baton Rouge, LA, 1971.

BIBLIOGRAPHY

Thurber, K. J., and Masson, G. M., *Distributed Processor Communication Architecture*. Lexington, MA: Lexington, 1979.

Tippie, J. W., and Kulaga, J. E., Design considerations for a multiprocessor-based data acquisition system. *IEEE Transactions on Nuclear Science* (August 1979).

Toong, H. D., and Gupta, A., An architectural comparison of contemporary 16-bit microprocessors. *IEEE Micro* (May 1981).

Torrero, E. A., Focus on microprocessors. *Electronic Design* (September 1974).

Torrero, E. A., Ed., *Microprocessors: New Directions for Designers*. Rochelle Park, NJ: Hayden, 1975.

Treybal, R. E., *Mass-Transfer Operations*. New York: McGraw-Hill, 1955.

Vacroux, A. G., Explore microcomputer I/O capabilities. *Electronic Design* (May 1975).

van Dam, A., Some implementation issues relating to data structures for interactive graphics. *International Journal of Computer and Information Sciences*, **1,** 4 (1972).

Vandenbrouche, L., Selecting CAD/CAM displays. In *CADCON West 84*, San Francisco, 1984. New York: Morgan-Grampian, 1984.

van den Bos, J., Definition and use of higher-level graphics input tools. In SIGGRAPH '78 Proceedings, published as *Computer Graphics*, **12** (3) (August 1978).

van Din, P., A draftsman's interface to solid modeling. In *CADCON East 84*, Boston, 1984. New York: Morgan-Grampian, 1984.

Van Winkle, M., *Distillation*. New York: McGraw-Hill, 1967.

Voelcker, H. B., and Requicha, A. G., Geometric modelling of mechanical parts and processes. *Computer* (December 1977).

Wallace, V. L., The semantics of graphic input devices. In Proceedings, SIGGRAPH/SIGPLAN Conference on Graphics Languages, published as *Computer Graphics*, **10,** 1 (April 1976).

Warner, J. R., Device-independent tool systems. *Computer Graphics World* (February 1984).

Warnock, J., The display of characters using grey level sample arrays. In SIGGRAPH '80 Proceedings, published as *Computer Graphics*, **14,** 3 (July 1980).

Wasserman, G., *Color Vision: An Historical Introduction*. New York: Wiley, 1978.

Wegner, P., Programming with Ada: An introduction by means of graduated examples. Englewood Cliffs, NJ: Prentice-Hall, 1980.

Wegner, W., Ed., *Research Directions in Software Technology*. Cambridge, MA: MIT, 1978.

Weisberg, D. E., Performance and productivity in CAD. *Computer Graphics World* (June 1983).

Weiss, B., Evaluating graph and chart output. *Computer Graphics World* (February 1984).

Weller, D., and Williams, R., Graphic and relational data base support for problem solving. In SIGGRAPH '76 Proceedings, published as *Computer Graphics*, **10,** 2 (Summer 1976).

West, J., Emerging importance of engineering workstations in manufacturing. In *CADCON East 84*, Boston, 1984. New York: Morgan-Grampian, 1984.

Westerhoff, T., Software in the future factory. *Test & Measurements World* (December 1983).

Wightman, E. J., *Instrumentation in Process Control*. Woburn, MA: Butterworth, 1972.

Wilde, D. J., *Optimum Seeking Methods*. Englewood Cliffs, NJ: Prentice-Hall, 1964.

Williams, R., On the application of relational data structures in computer graphics. In *Proceedings, 1974 IFIP Congress*. Amsterdam: North-Holland, 1974.

Wolf, S., *Guide to Electronic Measurements and Laboratory Practice*. Englewood Cliffs, NJ: Prentice-Hall, 1973.

Woodruff, G., Automated inspection of PWB's. Presented at Automated Manufacturing 1984, Greenville, SC.

Wong, E., and Youssefi, K., Decomposition—A strategy for query processing. *ACM Transactions on Database Systems* (September 1976).

Wu, S. C., Abel, J. F., and Greenburg, D. P., An interactive computer graphics approach to surface representation. *Communications of the ACM*, **20,** 19 (October 1977).

Yourdon, E., and Constantine, L. L., *Structured Design*. New York: Yourdon, 1975.

Zaks, R., *Microprocessors*. Berkeley, CA: Sybex, 1979.

Zambuto, D. A., Robotic system needs for non-standard components assembly in electronics manufacturing. Presented at Automated Manufacturing, 1984, Greenville, SC.

Zilog Corp. *Z8000 User's Guide*. Cupertino, CA: Zilog Corp.

Zimmerman, H. H., and Sovereign, M. G., *Quantitative Models for Production Management*. Englewood Cliffs, NJ: Prentice-Hall, 1974.

INDEX

INDEX

absolute position encoder, 189
absolute pressure, 50
absorbometer, 203
accelerometer sensitivity, 199
accounting-grade measurements, 150
accuracy, 2, 11
acoustic emission, 216
acoustic noise, 218
ADA, 274
ADP, 55
Advanced Data Communications Control Procedure (ADCCP), 279
agitator power viscometer, 133
air bubbler, 152
alternating-current permanent magnet tachometers, 193
aluminum oxide hygrometer, 137
ambient pressure, 50
analog multiplexing, 22
analog sensor, 250
angular accelerometer, 196
angular position liquid density sensors, 95–96
ANSI American National Standards Institute, 274
antenna probe, 151
aperture delay, 24
API hydrometer degrees, 94
APT, 261
arbiter, 267
Archimedes' principle, 159
ATLAS, 274
automatic control, 14
automatic efflux-cup viscometer, 134
average deviation, 9

backscatter Doppler, 86
ball float, 164
Balling degrees, 94
ball liquid density sensor, 96–97
Baumé degrees, 94
beat frequency oscillator, 217
benchmark, 263
bonded gauge, 59
boundary layer flowmeter, 80, 82
bridge errors, 47
brittle coatings, 232

brix degrees, 94
broad-band pyrometer, 43
bubble emission, 214

cadmium sulfide photocells, 212
calibration curve, 7
calibration cycle, 7
cantilever beams, 233
capacitance liquid density meter, 97
capacitance probe, 154
capacitive acceleration transducers, 197
capacitive devices, 185
capacitive pressure transducers, 56–57
carbon resistors, 34
Carrier Sense Multiple Access (CSMA), 277
catalytic combustion, 214
centrifugal gas density sensor, 112
charge amplifier, 198
CIM, 283
cippoletti, 86
collision detection (CSMA/CD), 277
common-mode errors, 27
compensating resistor, 65
compensation loop, 35
computer-aided manufacturing (CAM), 241
computer automation, 242
computer-integrated manufacturing (CIM), 273
computer numerical control (CNC), 259
concentric-thin-plate square-edge orifice, 72
concentric viscometer, 131
condenser microphones, 219
conductivity, 205
conductivity switch, 156
congestion-prone system, 280
continuous capillary viscometer, 129
continuous level detectors, 182
continuous processing, 249
control loop, 251
converters, 20
copper, 31
cross-axis sensitivity, 199
crosstalk, 23
curie, 168
curie point, 55

damped-sensor level switches, 181
dead band, 8

298 INDEX

demodulator, 187
density, 93-94
design automation, 243, 247
dew point, 135
dew-point elements, 136
dew point sensing, 140
diaphragm, 51
diaphragm switches, 156
differential pressure, 50, 157
diffused semiconductor strain gauges, 64
digital control systems, 191
digital interpolation, 254
digital pressure instruments, 67
direct-current generator/tachometers, 193
direct density controller, 112
direct numerical control (DNC), 250
direct weighing, 125
discrete-parts manufacturing, 249
displacement gas density sensor, 111-112
displacement liquid density sensor, 97-98
distributed control, 260
distributed systems, 254
Doppler effect, 85
droop, 24
dry bulb, 139
dual-slope integrating A/D converter, 20
Dunmore, 137
duplexed configuration, 266

electromagnetic flowmeters, 84-85
electromagnetic suspension density sensor, 99-100
electromechanical tachometer, 192
end effectors, 256
ETHERNET, 276-277
exponential velocity change, 255

factory automation, 241
factory automation networks, 268
factory information system, 273
fail-safe, 265
falling-piston viscometer, 133
Faraday's law, 84
fast neutrons, 47
feedback control, 15-16
feedforward control, 16
fiber optic system, 166
File Transfer Protocol (FTP), 281
fixed-coil velocity transducer, 192
flame rods, 213
flame-sprayed strain gauges, 226

flame spraying,
float and guide tube, 163
flow control, 280
flow nozzles, 74
flow rangeability, 72
fluid dynamic liquid density sensor, 100
flumes, 86
force-balance systems, 66
force-balance technique, 66
Fourier transforms, 257
frictional heating, 46

gamma rays, 167
gas density detectors, 95
gas purge, 153
gauge factor, 59, 224, 227
gauge pressure, 50
Geiger-Mueller, 170
Geiger tube, 104
G-M tube,
grounded beams, 235

hair hygrometer, 148
half-life, 167
head meters, 71
hierarchical network, 270
High Level Data Link Control (HDLC), 278
high-temperature strain sensing, 126
hot-wire anemometers, 80
hot-wire microphone, 221
hydraulic load cells, 116-118
hydraulic totalizers, 118
hydrometers, 101, 135
hydrophones, 223
hydrostatic head devices, 101
hygroscopic salt, 136
hysteresis, 2, 9
hysteresis errors, 51

ice-point references, 37
IC sensors, 36
IEE 488, 276
IEEE 802, 268, 278
IEEE 802/PROWAY, 276
impedance probe, 159
incremental encoder, 189
incremental optical encoders, 194
independent linearity, 9
inductance bridge sensors, 187
inductively coupled float, 176
inductive pressure transducers, 50-54

INDEX 299

inductive weight sensors, 122
industrial robots, 256
in-line torque sensors, 236
Instrumentation Society of America (ISA), 282
integrated factory system, 262
International Standardizing Organization (ISO), 275
Internet Protocol (IP), 281
ionic activity, 207
ionization chamber, 104

Jackson turbidity unit, 202

laser beam, 166
laser interferometer, 189
laser metal-cutting, 258
laser scribing, 259
laser welding, 258
layered architecture, 280
lead sulfide photocells, 212
level gauges, 164–165
Lindval glow microphone, 221
linear variable differential transformer, 54, 186
linear velocity transducers, 191
lithium chloride, 136
loop approach, 267
low-level multiplexers, 24
low-level signals, 253
LVDT, 54

magnetic displacement sensors, 188
magnetostrictive sensing techniques, 123–124
Mail Transfer Protocol (MTP), 281
manual programming, 260
manual teach programming, 258
Manufacturing Automation Protocol, 283
Manufacturing Message Format Standard (MMFS), 283
mass flowmeters, 89–91
mass spectrometer, 215
master processor, 265
master-slave network, 269
measured accuracy, 7
measurement, 1
mechanical hygrometer, 139
mechanical lever scales, 114–116
metal foil gauges, 225
Moiré fringe analysis, 233
monorail transducers, 125
motion balance controller, 109
motor control, 256

moving-coil microphones, 219
moving-coil transducers, 192
MRP, 283
multidrop configuration, 271
multiplexers, 19
multiplexing, 22–25

National Bureau of Standards (NBS), 281
natural frequency, 200
network, 265
Newtonian substance, 128
nickel, 32
Ni-Span C, 107
node, 269
noise, 25
non-Newtonian, 128
nonzero-based linearity, 9
nuclear radiation sensors, 121–122
Nuclear Regulatory Commission, 170
nucleonic flowmeters, 84

off-line programming, 258
offset error, 24, 229
omnidirectional microphone, 221
on-off transmitter level switches, 181–182
open-channel devices, 86
open-loop systems, 255
Open Systems Architecture, 275
Open Systems Interconnection, 275
optical dew-point hygrometer, 141
optical encoders, 189
optical level switches, 165–167
optical pyrometer, 41
organizational model, 245
orifice plates, 71–72
oscillating-Coriolis, 102
oscillating-vortex flowmeters, 80

Parshall flumes, 87
partial-radiation pyrometer, 43
peer-to-peer networks, 270
Peltier, 37
Peltier effect, 37
pendulum, 115
pH, 207
photoelasticity, 232
photoelectric encoders, 189
piezoelectric acceleration transducers, 197
piezoelectric energy, 4
piezoelectric flowmeters, 85
piezoelectric microphone, 220

300 INDEX

piezoelectric pressure transducers, 54–56
piezoelectric transducer, 234
piston meters, 88
piston pumps, 88
Pitot tube, 74
platinum film, 57
platinum resistance sensors, 32
pneumatic load cells, 118–119
pneumatic transmitter, 158
point detection, 150
point drift, 8, 9
point-to-point network, 272
Poisson's ratio, 226
Pope cell, 137
positive displacement meters, 88–89
potentiometric displacement transducer, 184
potentiometric transducers, 190
pressure drop techniques, 128
pressure-gradient microphones, 221
process automation, 247
programmable controllers, (PCs), 250, 255
proving rings, 233
psychrometers, 136
psychrometry, 135
p-type semiconductor, 228
Purdue Industrial Control Workshop (PICW), 282

quadrant-edge orifice plate, 72
quartz-crystal transducers, 56
Quevenne degrees, 94

radiation level sensors, 167–173
radiation liquid density sensor, 103–105
radiation pyrometers, 41
radiation techniques, 40–44
Rayleigh disk, 221
rectangular notch weir, 86
rectifying phototube, 211
redundancy, 266
relative humidity, 135
reluctance load cells, 122
remote multiplexing, 252
repeatability, 2
reproducibility, 2, 10
resistance strain gauge, 224
resistance tape detector, 173
resistance temperature detectors, 30–32
resistor pastes, 33
resolution, 2, 22
resonant frequency probe, 180

Reynolds number, 83, 129
ribbon microphones, 219
rolling-diaphragm hydraulic load cell, 116
rotameters, 77
rotary encoders, 188
rotating-cone viscometer, 132
rotating-paddle switches, 174–175
RS23C, 268
RS422, 268
RTDs, 30

sample hold circuits, 19
scintillation detector, 104
search coil, 217
Seebeck effect, 37
seismic mass, 197
self-heating effect, 45
self-heating error, 39, 45
self-learn software, 248
self-repairing, 266
self-test software, 248
semiconductor gauges, 227
semiconductor strain gauges, 121
semiconductor strain gauge transducers, 64
sensitivity, 2
sensor, 2, 4
shaft encoders, 188
shared-file, 266
shield, 27
signal conditioning, 19
Sikes, Richter Trailers degrees, 94
silicon, 228
silver-silver chloride couple, 209
simulation software, 248
single-float viscometer, 130
slewing, 24
slip rings, 238
slotted bending beam construction, 124
solid level gauging, 150
sound velocity liquid density sensor, 105–106
specific conductivity, 206
specific gravity, 93
spectroscopic hygrometer, 139
spring balance instrument, 162
spring balance scales, 114
star configuration, 270
stem conduction, 39
stem conduction errors, 46
stepping motor control, 255
straight-tube density transmitter, 110–111

INDEX

strain gauge load cells, 119
strain gauge pressure transducers, 59–65
strain gauge tachometer, 195
stress-strain gauge, 225
structured programming, 248
successive-approximation, 20
supervisory control function, 251
surface sensors, 178–179
switching angular speed tachometers, 195
Synchronous Data Link Control, (SDLC), 278–279
system accuracy, 12–14
system deadlock, 280

tape level devices, 175
Technical and Office Protocol (TOP), 283
temperature, 30
thermal conductivity, 204
thermal lay error, 39
thermal level sensors, 179
thermal radiation errors, 46
thermistors, 33–34
thermocouples, 36–40
thermography, 216
thin-film strain gauges, 63
Thomson effect, 37
tilt switch, 164
token passing, 266
toothed-rotor tachometer, 193
torque tube, 161
torsional vibration liquid density sensor, 106
total-radiation pyrometer, 42
transducer, 2, 4
transistors, 35
Transport Control Protocol (TCP), 281
trapezoid notch, 86
tungsten, 32
turbine flowmeters, 75–77

Twaddell degrees, 94
two-color pyrometer, 43
two-float viscometer, 130
Type 1 servo, 253

ultrasonic level detectors, 180–181
ultrasonic microphones, 221
ultrasonic viscometer, 131
unbonded gauge, 59
unit approach, 267
United Nations' Consultive Committee for Telephone and Telegraph (CCITT), 275
U-tube, 108
U-tube density sensor, 110
UV detector, 213

variable-area meters, 77–80
variable-reluctance transducers, 187
Venturi tubes, 73
vibrating-plate-type sensor, 106
vibrating-reed switches, 179–180
vibrating-reed-type viscometer, 132
vibrating-spool liquid density sensor, 107–108
vibrating-tube liquid density sensor, 108–109
vibrating-wire force transducers, 236
virtual address, 281
Virtual Terminal Protocol (VTP), 281
viscosity, 128
vortex shedding flowmeter, 83

weight-bulb-density sensor, 109
Weirs, 86
wet bulb, 139
Wheatstone bridge, 34
wire-guided thermal sensors, 177–178

X.25, 280

zero-based linearity, 10